# Moving Towards Low Carbon Mobility

**To our families**

# Moving Towards Low Carbon Mobility

*Edited by*

Moshe Givoni

*Department of Geography and Human Environment, Tel-Aviv University, Israel*

David Banister

*Oxford University, UK*

**Edward Elgar**

Cheltenham, UK • Northampton, MA, USA

Published by
Edward Elgar Publishing Limited
The Lypiatts
15 Lansdown Road
Cheltenham
Glos GL50 2JA
UK

Edward Elgar Publishing, Inc.
William Pratt House
9 Dewey Court
Northampton
Massachusetts 01060
USA

A catalogue record for this book
is available from the British Library

Library of Congress Control Number: 2012947336

This book is available electronically in the ElgarOnline.com
Social and Political Science Subject Collection, E-ISBN 978 1 78100 723 5

MIX
Paper from
responsible sources
FSC® C018575

ISBN 978 1 78100 722 8 (cased)

Typeset by Servis Filmsetting Ltd, Stockport, Cheshire
Printed by MPG PRINTGROUP, UK

# Contents

## PART II TOWARDS LOW CARBON MOBILITY

# Contributors

**Nihan Akyelken** holds a DPhil degree in Economic Geography from Oxford University and BSc (Economics and Philosophy) and MSc (European Political Economy) degrees from the London School of Economics and Political Science. Employing the principles of political economy, mobilities and economic geography, Akyelken has conducted research on public policy-making and futures research concerning globalization and logistics networks and political and economic geographies of infrastructure development.

**Malek Al-Chalabi** is a Doctoral Researcher at the Transport Studies Unit and Environmental Change Institute within the School of Geography and the Environment at Oxford University. His interests revolve around environmental citizenship, energy demand reduction and sustainable development and he has advised local and national governments on these issues.

**David Banister** is Professor of Transport Studies at Oxford University and Director of the Transport Studies Unit. Until 2006, he was Professor of Transport Planning at University College London. He has also been Research Fellow at the Warren Centre in the University of Sydney (2001–02) on the Sustainable Transport for a Sustainable City project, Visiting VSB Professor at the Tinbergen Institute in Amsterdam (1994–97) and Visiting Professor at the University of Bodenkultur in Vienna in 2007. He is currently the first Benelux BIVET-GIBET Transport Chair (2012–13). This is the 21st book that he has written or edited.

**Eda Beyazit** is a DPhil student at the Transport Studies Unit, Oxford University and also among the teaching staff of the Istanbul Technical University. In her research she investigates the distributional impacts of transport infrastructure investments. Beyazit holds BSc and MSc degrees in Urban and Regional Planning.

**Justin Bishop** is a James Martin Research Fellow in the Institute for Carbon and Energy Reduction in Transport (ICERT) of the Oxford Martin School. ICERT combines vehicle technologies, consumer preferences, novel transport fuels, resource limits, learning curves and policy to determine a realistic, low greenhouse gas-emitting passenger vehicle fleet.

**Moshe Givoni** is a Senior Lecturer at the Department of Geography and Human Environment, Tel-Aviv University, and a Visiting Research Associate at the Transport Studies Unit, Oxford University. He is also an Associate Editor for *Transport Reviews* journal. Givoni gained his PhD at the Bartlett School of Planning, University College London and received a Marie Curie postdoctoral fellowship which was undertaken at Vrije Universiteit (VU) Amsterdam. He was also a Research Fellow at Wolfson College, Oxford University.

**Robin Hickman** is a Senior Lecturer at the Bartlett School of Planning, University College London, and a Visiting Research Associate at the Transport Studies Unit, Oxford University. He is a specialist in transport and climate change, urban structure and travel, integrated transport and urban planning strategies, and the management of major multidisciplinary projects.

**Jian Liu** is a Researcher at the Energy Research Institute, National Development and Reform Commission of China. His research focuses on transport energy and environment, particularly on alternative fuel vehicles in terms of their life cycle energy consumption and emissions, infrastructure deployment strategies and future market potential. He graduated from Oxford University with a PhD in Geography and the Environment, the University of Newcastle upon Tyne (MSc) and Beijing University of Technology (BSc) in Transport Engineering.

**James Macmillen** is a Doctoral Candidate in the Department of City and Regional Planning, Cornell University, with research interests in planning history and theory. From 2008 to 2012 he was a Research Fellow at the Transport Studies Unit, Oxford University, where he remains a Visiting Fellow. Macmillen is a Fellow of the Royal Geographical Society and a 2012–13 US–UK Fulbright Scholar.

**Julia Markovich** is a Visiting Scholar at the School of Urban and Regional Planning (SURP), Queen's University. She has recently completed a scholarly manuscript on the professionalization of urban planning in Canada.

**Andre Neves** is a Landscape Architect and a Doctoral Researcher at the Transport Studies Unit, Oxford University, Ontario, Canada, focusing on the relationship between implementation of infrastructure for cyclists and pedestrians, changes in travel behaviour and associated carbon emissions.

**Tim Schwanen** is a Research Fellow at the Transport Studies Unit, School of Geography and the Environment, Oxford University. He obtained his PhD from Utrecht University in 2003 (cum laude). His diverse research interests include the relationships between transport and climate change as well as the geographies of mobilities, ageing and wellbeing. He has published over 70 contributions in international journals and from January 2013 is Editor-in-Chief of the *Journal of Transport Geography*.

**Martino Tran** is a Senior Oxford Martin Research Fellow and is interested in the dynamics of energy and environmental systems, focusing on technological innovation, decision theory and computational modelling. He holds a DPhil in Environmental Science and Applied Mathematics from Oxford University.

# Preface

This book celebrates 40 years of research at the Transport Studies Unit. It is the result of an extensive team effort at the TSU, Oxford University. It brings together the results of what started as a review of 'The science of mobility' for the Smith School of Enterprise and the Environment, Oxford University. Besides the team of authors who have contributed to this book, we would like to acknowledge the contribution in various ways of many others, including those who participated in two workshops, reviewed the chapters for the book and made useful interventions while the book was in preparation. We would like to thank colleagues at the Smith School including Angela Wilkinson, Mick Blowfield, Alex Williams, Christian Carey, Aisling O'Sullivan Darcy and Bill Sharp. We also acknowledge the contribution of colleagues at the Transport Studies Unit including Karen Anderton, John Bates, David Bonilla, Christian Brand, Tim Jones, Karen Lucas, Sally Pepperall, Gordon Stokes as well as others including Julian Allen, Casper Chorus, Torsten Fleischer, Sveinn Gudmundsson, Peter Hall, Sigal Kaplan, Susan Kenyon, Yael Parag, Paul Peeters, Antti Talvitie, Guihua Wang and Bert van Wee.

The Editors

**Moshe Givoni**
Department of Geography and the Human Environment
Tel-Aviv University

**David Banister**
Transport Studies Unit
School of Geography and the Environment
Oxford University

# 1. Mobility, transport and carbon

**Moshe Givoni and David Banister**

## INTRODUCTION

The wheel first came into use about 6000 years ago and since then it has been one of the most influential and useful inventions. Most personal movement is totally dependent on the wheel in some form or other, and it is an essential part of the ever moving world. Society is also reliant on the movements of other people and/or goods, and those movements taken together very much explain and define our society. In the not too distant past 'feet' shaped the boundaries of our perceived and experienced world, but now it is the wheel. The movement of people and goods has grown exponentially over the last 50 years, in both spatial scale and importantly in distance, and the world has been 'shrunk'. This has brought enormous benefits but some formidable challenges as well, some of which we are only now starting to realize, to understand and to act upon.

At the heart of the 'carbon problem of transport' is the two-way relationship between mobility and transport. These two concepts provide the links between our desire to move, to change our life situation, to fulfil our aspirations and to provide the means available to physically move. While the desire to move shapes our transportation system, it is at the same time shaped by it. Understanding the various linkages between mobility and transport is the basic and first step towards reducing the negative environmental impacts of transport, foremost among these being the need to reduce the carbon footprint of transport.

In many respects, most attention has been given to understanding the means by which transport impacts on the environment in various ways. Although this relationship is now relatively well understood, and leads to the logical conclusion that the seemingly unlimited desire to travel cannot continue to rise, it is difficult to observe any real progress towards reducing carbon dependent travel. The amounts of travel are continuing to increase, both at home and for long distances, and for both passenger and freight movements. One of the reasons for the lack of progress has been the relatively narrow perspective taken in understanding transport and in developing suitable policies to mitigate its carbon impact. In this book we look at transport and mobility as a system, and an investigation is made of each of the system's main elements before looking at them in an integrated, cross-cutting way, principally by addressing the carbon issue. On these pages carbon is used to represent more generally the environmental effects caused by the physical movement of people and goods, and the more general concern over the need to limit the total amount of non-renewable resources being used, and in turn to make the most efficient use of those resources that are needed.

## MOBILITY, TRANSPORT AND MOBILITIES

Today more people are travelling farther to reach places of employment, education, health, leisure and other activities. In the US people now travel 100 times further, in terms of distance travelled per person, than they used to travel in the 1800s (Buchanan, 2002). In France the average travelling distance per person per full day (excluding walking) stood at less than 100 m in 1800, rising to 1 km in 1900, and about 50 km towards the end of the twentieth century (Grübler, 2004). Similarly, travel distances in the UK have doubled over 40 years during the transition from the industrial to the post-industrial city (Banister et al., 2011). Modelling projections, Schäfer et al. (2009) have estimated that in 2050 the average American will travel 48 000 km per year (almost 130 km per day), and while the current world population travels a total of 23 billion km per year, it is estimated that this level will increase to 104 billion km in 2050 (Schäfer et al., 2009). It is not only that people are more mobile – freight transport has also increased substantially, as illustrated by the fact that the world international seaborne trade total has risen from 2500 million tonnes loaded in 1970 to almost 8500 million in 2010 (UNCTAD, 2011).

These trends demonstrate the profound changes that are taking place in mobility patterns, as measured by the travel volumes, and this is true across all modes of transport. But the main increases in mobility have come from the carbon intensive forms of transport, as faster, longer journeys have resulted in more energy being used and more emissions created. 'The whole world seems to be on the move', as Sheller and Urry (2006, p. 207) note, but it is not just the people and goods that are moving; ideas, information, culture, capital and money are also on the move.

'Mobility' relates to the total amount of travel that is undertaken on all forms of transport, and it includes the physical movement from one place to another. It covers both the movement of people and freight. When mobility is carried out by motorized forms of transport it results in the emissions of carbon and other pollutants. 'Transport' is broader than mobility and includes the modes of transport as well as the supply of transport through various institutions and organizations. Transport and mobility are thus closely related and linked, and it is for this reason that this book discusses the 'transport and mobility' system in a holistic way. 'Mobilities' is a more contested and ambiguous concept (see also Chapter 2). John Urry defines the 'mobilities paradigm' as 'a wide array of economic, social and political practices, infrastructures and ideologies that all involve, entail or curtail various kinds of movement of people, or ideas, or information, or objects' (Urry, 2007, p. 43). Mobilities are thus considered to be about movement as well as potential movement, and no-movement (blocked movement in Urry's words) of essentially everything, material, tangible or otherwise. As such, it does not necessarily result in carbon emissions, and does not always impact on the environment, as this happens when it is translated into actual mobility. Mobilities are one of the main explanatory factors for the demand for transport and for mobility. In all of the three concepts, transport, mobility and the mobilities paradigm, distance is central, 'as almost the key issue with which social life involving a complex mix of presence and absence has to treat' (Urry, 2007, p. 54). For physical movement to actually take place there is a need for another element to be present, namely, the transport infrastructure. Without infrastructure travel cannot take place. Infrastructure as such does not generate movement, as it only allows it to take place. Mobility can therefore be seen as being situated between the demand for transport and the infrastructure that allows this demand to be realized.

Having defined the concepts of mobility and mobilities, and placed them within the wider context of the changing patterns of travel, this introductory chapter now expands briefly on the carbon problem as it relates to transport. This is important as the underlying rationale for the book is to argue for low carbon mobility. This rationale also embraces some wider issues, such as the levels of resource use in transport, its dependence on one source of energy (oil), its role in allowing global patterns of trade and travel to develop and the more general need to live within the constraints of the planet now and in the future. The chapter also introduces the discussion of the differences between a high and low carbon mobility transport system, and briefly outlines the structure of the book.

## THE CARBON 'PROBLEM' OF TRANSPORT

Motorized transport is totally dependent on oil for its energy, and this has important implications, both for the availability and price of that oil and also for the emissions of carbon dioxide ($CO_2$). Transport accounts for about 24 per cent of global $CO_2$ emissions. It also accounts for over 61 per cent of global oil consumption (IEA, 2009), with more than 70 per cent of transport fuel consumption being used in the road sector. Transport needs to play a key role in contributing to global $CO_2$ reduction targets, but at present no reduction has taken place, as greater efficiency in the use of energy has been more than outweighed by the growth in travel. The increase in global $CO_2$ emissions from transport (excluding aviation and maritime transport) has grown by 32 per cent to 6.5 Gt $CO_2$ (1990–2009; IEA, 2011 and Table 1.1).

Global oil prices are at historically high levels and they are likely to remain high (over $120 per barrel for Brent crude – May 2012), as demand is strong and there is increasing instability in many of the major producing countries. This means that prices have risen substantially in those countries that allow the full costs to be passed on to the final user, but there are still considerable variations between countries as the taxation regimes are very different. Fuel in Europe costs about $1.80 per litre, nearly three times the price in the US. But it is in the developing nations of Asia that the growth in vehicle ownership and the demand for fuel is now taking place. Added to this growth is the problem that in many developing countries fuel is heavily subsidized, so that the incentives for drivers to use less fuel is reduced, and governments are using valuable foreign currency earnings to support motorization, meaning that less resources are available for public expenditure elsewhere in the economy.

Fuel security and high oil prices are a key concern to all governments, as production and prices are affected, which in turn impacts on global trade, economic performance and human welfare. There are also likely to be second round effects as the costs of fertilizers, food processing and the transport of goods will rise, which in turn leads to higher food and energy costs, often impacting more severely on low income families.

In addition to the total dependency of motorized transport on one source of energy, there are the implications for emissions, and in particular $CO_2$. Although there are many pollutants resulting from the use of oil, many can now be filtered out through catalytic converters and particulate traps, and this means that in many ways the levels of pollution from motorized transport are much less than in the past. But there is no technology that can currently reduce the levels of $CO_2$ emitted from burning fuel, and every kilogram of diesel or petrol produces about 3.16 kg of $CO_2$. Globally, there are tremendous differences in the levels of $CO_2$ emissions and in those emitted from transport (Table 1.1).

Globally, $CO_2$ emissions are 4.29 $tCO_2$ per capita, with the figure for transport accounting for about 1 $tCO_2$. The two figures for the Americas

*Table 1.1 $CO_2$ emissions levels overall and for transport (2009)*

| 2009 | $CO_2$ emissions $MtCO_2$ | $CO_2$ emissions ($tCO_2$) per capita | $CO_2$ transport emissions $MtCO_2$ | $CO_2$ transport ($tCO_2$) emissions per capita |
|---|---|---|---|---|
| World | 28999 | 4.29 | 6544 | 0.968 |
| OECD[a] | | | | |
| Americas | 6180 | 13.27 | 1940 | 4.166 |
| Asia Oceania | 2099 | 10.00 | 418 | 1.996 |
| Europe | 3765 | 6.85 | 957 | 1.742 |
| Non-OECD | | | | |
| Africa | 927 | 0.92 | 233 | 0.230 |
| Asia | 3153 | 1.43 | 492 | 0.223 |
| China | 6877 | 5.14 | 476 | 0.356 |
| Middle East | 1509 | 7.76 | 329 | 1.689 |
| Europe/Eurasia | 2497 | 7.46 | 346 | 1.032 |
| Latin America | 974 | 2.16 | 339 | 0.751 |

*Note:* a. Organisation for Economic Co-operation and Development.

*Source:* IEA (2011).

are three times higher overall and more than four times the average transport level, while the corresponding figures for Europe are 1.6 times and 1.7 times the global averages. Even in China the overall figure is 20 per cent higher than the global average, but the transport figure is lower at only 36 per cent of the average. It is clear that the developed countries must substantially reduce their carbon emissions, both overall and for transport. But the reality is that all countries are increasing their emissions and thus raising the global averages both overall and particularly for transport.

The carbon debate has only really become a dominant issue in both scientific research and political terms since 1997, when there was international agreement that the developed countries[1] should reduce their emissions of $CO_2$ overall by 5.2 per cent (1990–2012). Yet over the intervening period transport has made no contribution to those reduction targets. Overall there has been some progress in reducing $CO_2$ emissions, as levels have stabilized in the EU27 at about 4400 $MtCO_2$ (a reduction of less than 4 per cent on 1990 levels), but the levels for transport have continued to increase from 941 $MtCO_2$ (1990) to 1257 $MtCO_2$ (2008). This means that the transport proportion of the total has increased from 21 per cent of the total EU27 $CO_2$ emissions to 29 per cent. In addition, there

*Table 1.2 Travel per person in the EU27*

| Travel per person EU27 | 1990 | 2000 | 2005 | 2025 | 2035 | 2050 |
|---|---|---|---|---|---|---|
| City and < 150 km | 9000 | 9700 | 10000 | 12400 | 13000 | 14900 |
| Long distance > 150 km | 1500 | 1600 | 1650 | 2100 | 2200 | 2500 |
| Air | 500 | 670 | 760 | 1370 | 1620 | 2100 |
| Freight | 3800 | 4500 | 4900 | 6750 | 7400 | 8600 |
| Sea | 2500 | 2800 | 3100 | 4000 | 4500 | 5200 |

*Note:* Long distance is over 150 km and all distances are given as averages per person in the EU27. The passenger figures are in km per person per year and freight figures are tonne km per person per year. The top three rows are in kilometres per person and the bottom two are tonne km per person. Freight in the European Technology Assessment Group (ETAG) report assumes 80 per cent of tkm is transported over distances >150 km, so the figures have been grossed up and the maritime sector added, but it only covers internal and intra EU movements. For passenger travel, aviation is separated and rail is divided 15 per cent long distance and 85 per cent short distance, with the same division being used for car and motor cycles. Public transport has been added to city transport. International aviation and maritime transport outside the EU27 are not included.

*Source:* Based on authors' calculations and data from ETAG (2008) and EU (2011).

are the global emissions from maritime and aviation, both of which fall outside the Kyoto agreement (EU, 2011).

Much of the research on transport and climate change has concentrated on the city (for example, Banister, 2005), but this market is relatively stable and there are good opportunities here for moving towards more efficient low carbon transport systems. It is in the long distance market that substantial increases in travel are taking place, and this is also where there are far fewer opportunities available for reductions in carbon based energy use.

Travel in the EU27 may be divided into five mutually exclusive forms of travel, one covering short distance city passenger travel (< 150 km), a second covering long distance passenger travel (> 150 km), and others for air travel, sea and freight. Figures for the EU27 illustrate the scale of the problem (Table 1.2). City travel accounts for more than 98 per cent of passenger trips and about 80 per cent of distance (1990). Long distance passenger travel and air travel in the EU27 countries account for 20 per cent of distance, but only 2 per cent of trips. It is in these non-urban travel markets that the low carbon alternatives are much harder to envisage, let alone implement.

The main point to make here is the scale of change needed to reduce travel levels by 80 per cent in 2050. This is the statutory target that has been set in the UK (Committee on Climate Change, 2008). Globally, there is a need to reduce carbon emissions by 50 per cent (on 1990 levels) if global warming levels are to be limited to an increase of 2°C (Anderson and Bows, 2011). If no increase in efficiency was achieved, then distance would have to be reduced to 2400 km (on 2000 levels), rather than the forecast level of 19 500 km – this is more than 85 per cent reduction (Factor 8). The corresponding freight figures would be 1460 km (on 2000 levels), rather than the forecast level of 13 700 tkm – this is more than a Factor 9 (Committee on Climate Change, 2008). The key question is how far technology can help in achieving such a change, and how much would depend on travelling less and relying on the most efficient forms of transport.

It is commonly stated that reductions in transport are difficult and expensive, but there are many options available and this is the core argument underlying this book, namely, that transport has taken a low profile when it comes to making a substantial contribution to energy and carbon reduction targets. Its dependence on oil and importance in terms of facilitating accessibility and global trade have always dominated the debate, suggesting that a less resource consumptive transport sector would potentially reduce economic growth. The arguments presented here suggest that this is not the case, and that a leaner, more efficient transport sector would benefit the economy and reduce reliance on imported fuels. This situation is presented in the next two sections where the high and low carbon mobility options are elaborated.

## HIGH CARBON MOBILITY

The net result of the dominance of economic growth, global trade and efficient transport in the debate has been high levels of mobility that translate to high volumes (and distances) needed for transport. This in turn leads to high carbon mobility. While contemporary mobility practices arise through the interaction of numerous sociological, cultural, political and economic processes, a complexity that is vital to appreciate, the physical relationship between transport and carbon emissions is straightforward and it is governed by three fundamental variables: (1) the amount of transport (2) the amount of energy required per unit of transport and the distance travelled. This is the energy intensity of transport; and (3) the level of carbon emission (and other gases/impacts) released per unit of energy consumed.

Traditionally, much transport investment has been predicated on the lack of capacity in the system and the belief that congestion could be 'solved' by additional capacity. Investment decisions were justified on the basis of travel time savings and so the net effect was faster (and longer) journeys that used more energy. The high mobility option places a higher value on travel speed and reducing travel times, and this has led to longer travel distances. It has also resulted in a greater dependence on motorized transport and an increase in the use of energy and carbon. Recently, more attention has been paid to traffic management and demand management, but substantial investment in infrastructure continues in many locations.

These problems are at the centre of contemporary transport policy. Efforts to reduce carbon emission from transport largely fall under three categories: avoid, shift and improve (Schipper et al., 2000; Dalkmann and Brannigan, 2007). 'Avoid' suggests reducing the demand for transport and thus reducing its volume. Some volume of transport will always be required and this volume therefore needs to be 'shifted' to the most environmentally efficient modes. Furthermore, the environmental performance of these modes needs to be further 'improved', so that carbon (and other emissions) per unit of necessary transport by the most efficient mode of transport will be further reduced and minimized. For various reasons this strategy, while sound, has not yet resulted in the desired outcome. These reasons are explored throughout the book (see also Banister et al., 2011) and can be explained by current norms, centred very much around the meaning, and perceived value, of time, and the central tenet of transport policy thinking that travel is a derived demand. This is the current transport planning paradigm. The car is central to the current mobility and transport system, and considerable attention is given to this fact throughout this book.

There are three overarching elements, or assumptions, from which the current transport planning paradigm is derived. First, it is often argued that economic growth (usually measured as the increase per year in gross domestic product (GDP) per capita) is paramount and therefore a major political concern, especially in times of economic crisis. Second, transport is seen not only as enabling economic growth but also as facilitator of it. The transport system therefore must allow the smooth flow of people and goods and at increasing volume, 'bottlenecks' in the transport system must be eliminated. When this is not the case, and congestion arises, it is a cost to the economy. This cost is high since in the current mobility system 'time is money' and therefore faster is better (Banister, 2011). This is the third element, and this value of time is closely related to notions of consumerism and materialism, where more and faster are assumed better.

In many parts of the transport system demand exceeds the capacity of the transport networks, leading to major congestion problem in those bottlenecks. Forecasts for economic growth translate into forecasts for increased demand for transport and concern that lack of investment in increasing capacity on the transport network will result in more congestion, thus an economic burden, and this will hamper the prospects for economic growth. This has been the prevailing rationale for the 'predict and provide' approach, where demand is predicted and then the infrastructure has been provided. One of the outcomes of this approach is the generation of induced demand (Goodwin, 1996), and this is one of the reasons why bottlenecks always remained in the transport system, and congestion has more often increased than decreased. 'We cannot build our way out of congestion' is a saying associated with the recognition that there is a need for an alternative to the predict and provide approach.

This alternative has been reinforced by the recognition of the adverse impacts of transport on the environment. While notions of 'sustainable transport' and 'transport demand management' as alternatives to 'predict and provide' are increasingly considered in the transport planning debate (Owens, 1995; Banister, 2008), investment in transport infrastructure to support economic growth and reduce congestion is still the dominant factor in contemporary transport planning. Although the debate is changing emphasis, the transport system is still very carbon intensive. The high carbon mobility is centred on the private car, in which movement is embedded in many social norms or results from them, and the value placed on speed and the increasing amounts of travel.

## LOW CARBON MOBILITY

Low carbon mobility is defined as mobility that results in substantially lower levels of carbon. There are three ways of doing this, either individually or in combination: the first is to change social norms to ones that require and imply less mobility; the second is to follow a different economic growth model, namely, one that is not coupled with transport volumes; and the third is to switch to a different transport system, namely, one that results in less carbon. These options are defined and explored in detail in the second part of this book, which brings together all of the themes discussed in three different ways. There is a chapter that explores the alternative pathways to a low carbon mobility society (Chapter 13). This is followed by a more theoretical chapter that discusses the means by which change, from high to low carbon mobility, can take place through transition theory (Chapter 14). Chapter 15 outlines two different perspectives on city transport in a post carbon society, one based on more use of walking and cycling, and the other on clean technology (Chapter 15). Part II of the book sets in motion the process of 'thinking change', emphasizing the importance of action now, and these important issues are discussed in more detail in the concluding chapter that also acts as a summary to the book (Chapter 16).

However, a meaningful discussion of low carbon mobility must build on a comprehensive understanding of the current system of high carbon mobility. This is the objective of the first part of the book, where attention is given in turn to ten key elements of the transport system. These are: mobility cultures (Chapter 2), accessibility and equity (Chapter 3), governance and policy (Chapter 4), urbanization (Chapter 5), travel behaviour (Chapter 6), transport technology (Chapter 7), energy mix (Chapter 8), finance and investment (Chapter 9), supply chains (Chapter 10) and finally transport and the environment (Chapter 11). These elements are also key individual focus points in their own right, as they provide a more complete understanding of the interconnectedness of transport research. Yet, it is only when they are considered together that the transport system can be envisioned in its totality, and this is presented in 'Mobility as a complex system' (Chapter 12).

Thinking about low carbon mobility is not new, and substantial research and policy debates have been initiated in trying to move in this direction, but so far with only limited success. Too much focus has been placed on the actual detailed planning and design of the transport system. Increasingly, and importantly, new interpretations have begun focusing on understanding travel behaviour, and culture and contributions from various disciplines, but there is often not enough attention given to how all the elements of the transport system work together. Furthermore, in trying to deal with the low carbon mobility challenge, there is only limited understanding of the mutual relations and interdependencies between mobility and mobilities.

This book tries to address these important social, cultural and behavioural issues, as well as the more conventional economic and technological dimensions, so that a more complete understanding of the problems can be made, together with a strong case for a range of large scale interventions in the transport sector. The book provides a cohesive picture on what is happening and what needs to change in the mobility and transport system. The main message is that the complexity of the system should not encourage inaction, or business as usual or weak action. The book advocates the need for systemic change and for implementing substantial and immediate action.

# NOTE

1. These are the 40 Annex 1 countries, as defined in the Kyoto Protocol and this came into operation on 15 February 2005.

# REFERENCES

Anderson, K. and A. Bows (2011), 'Beyond "dangerous" climate change: emission scenarios for a new world', *Philosophical Transactions of the Royal Society A*, **369**, 20–44.

Banister, D. (2005), *Unsustainable Transport: City Transport in the New Century*, London: Routledge.

Banister, D. (2008), 'The sustainable mobility paradigm', *Transport Policy*, **15**, 73–80.

Banister, D. (2011), 'The trilogy of distance, speed and time', *Journal of Transport Geography*, **19**, 950–9.

Banister, D., K. Anderton, D. Bonilla, M. Givoni and T. Schwanen (2011), 'Transport and the environment', *Annual Review of Environment and Resources*, **36**, 247–70.

Buchanan, M. (2002), *Nexus: Small Worlds and the Groundbreaking Science of Networks*, London: W.W. Norton.

Committee on Climate Change (2008), *Building a Low Carbon Economy – The UK's Contribution to Tackling Climate Change*, December, Norwich: HMSO.

Dalkmann, H. and C. Brannigan (2007), *Transport and Climate Change – Sustainable Transport: A Sourcebook for Policy-makers in Developing Cities*, Eschborn: GTZ.

ETAG (European Technology Assessment Group) (2008), 'The future of European long distance transport', Deliverable 5, August, European Technology Assessment Group.

EU (European Union) (2011), *EU Energy and Transport in Figures, Statistical Pocketbook 2009/2010*, Brussels: Eurostat, DGTREN.

Goodwin, P.B. (1996), 'Empirical evidence on induced traffic: a review and synthesis', *Transportation*, **23**, 35–54.

Grübler, A. (2004), *Technology and Global Change*, Cambridge: Cambridge University Press.

IEA (International Energy Agency) (2009), *Transport, Energy and $CO_2$: Moving Towards Sustainability*, Paris: IEA, available at http://www.iea.org/textbase/nppdf/free/2009/transport2009.pdf (accessed 5 October 2011).

IEA (International Energy Agency) (2011), *$CO_2$ Emissions from Fuel Combustion – Highlights*, Paris: IEA, available at http://www.iea.org/co2highlights/co2highlights.pdf (accessed 9 March 2012).

Owens, S. (1995), 'From "predict and provide" to "predict and prevent"? Pricing and planning in transport policy', *Transport Policy*, **2**, 43–9.

Schäfer, A., J.B. Heywood, H.D. Jacoby and I.A. Waitz (2009), *Transportation in a Climate-constrained World*, Cambridge, MA: MIT Press.

Schipper, L., M.-L. Celine and R. Gorham (2000), 'Flexing the link between transport and greenhouse gas emissions: a path for the World Bank', June,

Washington, DC: World Bank, available at http://www.iea.org/textbase/nppdf/free/2000/flex2000.pdf (accessed 23 June 2012).
Sheller, M. and J. Urry (2006), 'The new mobilities paradigm', *Environment and Planning A*, **38**, 207–26.
UNCTAD (United Nations Conference on Trade and Development) (2011), *Review of Maritime Transport*, Geneva: United Nations, available at http://www.unctad.org/en/docs/rmt2011ch1_en.pdf (accessed 23 June 2012).
Urry, J. (2007), *Mobilities*, Cambridge: Polity Press.

# PART I

# Key elements in the mobility and transport system

# 2. Mobility cultures

**Eda Beyazit**

The start of the twenty-first century is marked by high levels of mobility: people, capital, finance, technology, commodities, ideas and information as well as risks and disasters are increasingly recognized as being 'on the move'. As noted in Chapter 1, more people are travelling further to reach places of employment, education, health, leisure and other activities and there is a general consensus on the need for reducing carbon dioxide ($CO_2$) emissions, especially from transport. Strategies are being developed by governments and especially by companies in the energy and automobile sector in order to overcome the consequences of this situation. However, the main criticism towards this approach is that such measures solely cannot reduce the negative impacts of transport on climate, and more drastic changes are required. Banister (2008) proposes the sustainable mobility paradigm as a way forward to solving the problems that we are facing, arguing that there is a crucial need for a change especially in the way in which people travel and how much they travel.

In the above context, the purpose of this chapter is to understand the role of the culture of performing mobilities in influencing the way in which people travel and how much they travel, and to find out whether there is a need to reformulate our 'mobility cultures' in order to build a low carbon society. First, the new mobilities paradigm is briefly reviewed. This is followed by a discussion on how the mobilities are practised and finally, the meaning of these practices for a low carbon society is evaluated.

## THE NEW MOBILITIES PARADIGM

> It is good to have an end to journey towards; but it is the journey that matters in the end. (Le Guin, 1969)

In the last decade a vast amount of research has been carried out in various disciplines including anthropology, cultural studies, geography, migration studies, science and technology studies, tourism and transport studies, and sociology that have helped to form the new mobilities paradigm, a term that was introduced by Mimi Sheller and John Urry (2006). Although Cresswell (2011) points out the importance of acknowledging previous engagements in the literature with different dimensions of geographies of mobility; the paradigm, as we understand it now challenges the ways in which social science research has been 'a-mobile' and has failed to recognize the significance of an increasingly mobile and interconnected global society.

What is making the new mobility paradigm so interesting for many scholars is the development of a new sociological understanding of movement that crosses disciplinary boundaries. It is arguably a more holistic investigation of movement in a variety of areas free from disciplinary and sub-disciplinary boundaries (Cresswell, 2010a). Although it is possible to state that the mobilities research has primarily been a product of sociology (Urry, 2000, 2007a, 2007b), it has been developed through the contribution of many disciplines. Geographers and academics from related disciplines have placed people and things that are on the move in the centre of the discipline (Cresswell and Merriman, 2011). Different approaches to these mobile entities have helped in developing several theories such as the migration theory in population and development geography, the tradition of time-space geographies, accounts of journeys to work developed within feminist geography, geographies of tourism, choreographies of place developed within the phenomenological tradition, historical geographies of transport and mobility, geographies of commodities, globalization and capital, research on the imaginative geographies of travel writing and accounts of the role of travel and exploration (Cresswell and Merriman, 2011, p. 4).

One might question what it is then that the new mobilities paradigm suggests that has not yet been discussed within the social sciences. The new mobility paradigm proposes a more critical way of thinking about the formation of the social world. Cresswell (2010a, p. 18) discusses that 'the [new] "mobilities" approach brings together a diverse array of forms of movement across scales ranging from the body (or, indeed parts of the body) to the globe'. The new mobility approach focuses on people and the meanings they attach to social phenomena while on the move. Therefore, it departs from, for instance, functional rationalist approaches that have dominated transport research with its preoccupation with cost and time of moving between points A and B devoid of significance or meaning and typically constituted as 'dead time'.

The new mobilities paradigm then draws our attention to the practices, spaces, subjects, meanings and the politics while on the move (Adey, 2010; Cresswell and Meriman, 2011) or while staying still (Cresswell, 2011). Spaces at rest such as airports, car parks and ports, and spaces that enact mobility such as roads, railways and bridges are needed for mobilities (Cresswell and Meriman, 2011). 'These spaces produce structural or infrastructural contexts for the practicing of mobility. They are agents in the production of mobilities' (Cresswell, 2011, p.7). Likewise, practices of mobility produce spaces. Therefore, it is possible to mention a mutual relationship between these two notions. The third dimension of this relationship is the meanings and how these meanings are associated with the movement. The fourth dimension, the subjects, may require a relatively more complicated debate as their relationship in the mobilities debate is not only formed through their relationship with spaces, practices and meanings but also with their 'subject positions' (Cresswell, 2011). These positions are represented under different groups such as gender, age, ethnicity, wealth, class, nationality and abilities. It is therefore this stratification within society that regulates different experiences of mobilities and leads to a discussion on the politics of mobility (Adey, 2010). As Cresswell (2010a, p. 20) discusses, 'forms of mobility (walking, driving, etc) and these aspects of mobilities (movement, representation, and practice) are political – they are implicated in the production of power and relations of domination'. The rest of this chapter is devoted to understanding the term mobility cultures and different forms of mobility.

## MOBILITIES AS CULTURE: PERFORMING MOBILITIES

> . . . the performing of mobilities is culture. (Jensen, 2009a, p. xv)

The main subject of the new mobilities paradigm is people and their relationships with space, meanings, objects and images through the act of movement. As Jensen (2009a, p. xv) discusses, 'even if we study the physical movement of objects like people, cars, bikes, or goods we are simultaneously dealing with social issues of norms, power, identity and *culture formation*. It is fairly evident that mobility as a specific social practice may be related to cultural norms and regulations' (emphasis added; see also Jensen, 2009b). Therefore, it is important to understand what are these norms, power and identities and the formation of mobility cultures.

Mobility norms within localities are important factors that shape the culture of mobility. These norms may well be present in everyday life, from how to wait in the queue for the bus to how to behave in a public transport vehicle. These mobility practices are determined by societies themselves with or without being attached to governmental regulations and may be named as different 'codes' of mobility (Jensen, 2009a). These codes are taught by families to their children, starting from the first steps of a child in traffic and include basic knowledge such as which direction to look first while crossing streets or how to wait at traffic signals. However, as we grow older and use different modes of transport such as buses, trains and planes we learn mobility codes for these modes and develop our own 'mobility knowledge' (Jensen, 2009a). Walking codes, cycling codes, driving codes and so on differ from one society to another and such 'mobility cultures are linked to official and legal sanctions and mobility regulations' (Jensen, 2007, p. 3).When we visit other countries with different mobility codes we find ourselves eventually adapting to the mobility culture of those countries. Thus, developing mobility knowledge is an ongoing process throughout our lives. Likewise, new codes introduced by states identify the way we travel. For instance, rules at airports are quite determinant in the ways that we pack our suitcases, behave at the airports, avoid certain discussion topics and time our journey accordingly.

Power can be materialized in forms of capital, gender, race, physical wellbeing and capabilities, therefore, it is at the very heart of the mobility debate. Individuals or groups of individuals experience mobility in different ways and levels depending on their position in society. Therefore, as Massey (1994) highlights, power is an important component in shaping mobility among different social groups:

> different social groups, and different individuals, are placed in very distinct ways in relation to these flows and interconnections. This point concerns not merely the issue of who moves and who doesn't, although that is an important element of it; it is also about power in relation to the flows and the movement. Different social groups have distinct relationships to this anyway differentiated mobility: some people are more in charge of it than others; some initiate flows and movement, others don't; some are more on the receiving-end of it than others; some are effectively imprisoned by it. (Massey, 1994, p. 149)

Massey (1994) also states that although disadvantaged groups contribute to increasing mobility levels, they are not necessarily the ones who are benefiting from this contribution in the end. Moreover, she argues that unequal mobilities is not only a question of unequal distribution of mobility between different groups, but it is also relational, meaning that some groups can actually weaken other people's ability to be mobile (Massey, 1994, p. 150). Other dimensions of power asymmetry have been highlighted by Sheller and Urry (2006, p. 213) who state, 'moving between places physically or virtually can be a source of status and power'. Likewise, Skeggs (2004, p. 49) mentions that 'mobility and control over mobility both reflect and reinforce power. Mobility is a resource to which not everyone has an equal relationship.' Drawing attention to the consequences of these power relationships, Manderscheid (2009) discusses how unequal power relations that are generated by different subject positions may create 'unequal mobilities'.

Issues related to age, wealth, ethnicity and physical wellbeing, and their relationship with lack of mobility have been widely discussed in the transport geography literature, mainly in the social exclusion debate (Church and Frost, 1999; Social Exclusion Unit, 2003; Lucas, 2004; Rajé, 2004). However, a vast amount of research has been carried out within mobilities research focusing on gendered mobilities, mainly by feminist geographers. As Cresswell (2010b, p. 554) points out, no group other than feminist geographers has been 'more attentive to the variety of mobile experiences and the power dynamics afforded by mobility'. Elsewhere, Uteng (2009) and Uteng and Cresswell (2008, p. 2) point out 'how people move (where, how fast, how often etc.) is demonstrably gendered and continues to reproduce gendered power hierarchies'. Therefore, it requires particular attention.

Societies where women's role is more narrowly defined can serve to define the limits of their mobility. Early discussions of mobility and gender can be found in Massey's work where she discusses that there is a limit to women's mobility: 'women's mobility, for instance, is restricted – in a thousand different ways, from physical violence to being ogled at or made to feel quite simply "out of place" – not by "capital", but by men' (Massey, 1994, p. 148). In societies where women's role is less narrowly defined their mobility landscape can be more complex, for example, accomplishing daily routines that can typically include juggling domestic duties, organizing child care in addition to employment. Therefore, women may be restrained from their freedom of movement.

As seen in a striking late nineteenth-century example of 53-year-old Frances Willard learning to ride a bike, examined by Hanson (2010, p. 5), rediscovering freedom of movement may help build up women's confidence and character, and may act as a means of challenging male dominance, improving prevailing norms of masculinity and as an impetus for transforming gender relations. Research shows that there are significant differences between the daily travel patterns of women and men, which for the former are far more complicated and include shorter trips and more women use public transport (Uteng and Cresswell, 2008). In a different study exploring the differences in the mobility patterns between Norwegian and non-Western immigrants Uteng (2009) discovered that non-Western women immigrants spend longer hours travelling compared to their male counterparts and Norwegian women immigrants due to lack of public transport after certain hours and lack of car ownership. Moreover, the immigrant women in Norway have significantly less access to cars than the immigrant men.

The formation of mobility cultures is affected by local, national and international regulations. While globalization promises seamless travel between countries, in reality state politics control mobility between states. As Sheller and Urry (2006, p. 210) highlight, '[r]ights to travel are highly uneven and skewed even between a pair of countries'. Airport regulations or country entrance visas largely change our perception of travelling and they often put many people off travelling at all.

## MOBILITY PRACTICES

> Our mobilities create spaces and stories – spatial stories. (Cresswell and Meriman, 2011, p. 5)

Mobility cultures are not only formed by cultural norms and social constraints. In addition, the ways in which the mobilities are practised gain importance. These embodied practices can be experienced in different ways, and they are 'central to how we experience the world, from practices of writing and sensing to walking and driving' (Cresswell and Meriman, 2011, p. 5). Mobility is a socio-spatial experience and the method of movement shapes that experience. This section provides examples of experiencing mobilities through different transport modes as the spaces where these spatial stories are created. Modal choice can be regarded as an essential way to claim the freedom that is provided by the act of movement (Petersen, 2007). Mobilities can be experienced in two main ways: by personally performing them (as in walking, cycling and driving) and by moving with others (Bissell, 2009a). In order to understand the former type of performing mobilities, an important term, automobility is briefly explained.

The significance of notions of individual freedom in elevating the motor car to the dominant mode of conveyance in industrialized societies is an important case to point out:

> The term automobility works off the combination of autonomy, and mobility. In its broadest sense we can think of many automobilities – modes of autonomous, self-directed movement. . . . The promise here is for self-steering autonomy and capacity to search out the open road or off-road, encapsulated in vehicles which afford not only speed and mobility, but act as comforting protected and enclosed private spaces, increasingly a platform for communications media, that can be enjoyed alone or in the company of significant others. (Featherstone, 2004, pp. 1–2)

The act of driving, the relationships that the driver establish with his or her environment, that is, with other passengers, pedestrians on the road, other cars, or the surrounding built environment or the nature, visual elements, music on the radio or other pleasant or unpleasant sounds, ideas and other elements of mobility construct the important daily interactions of a driver. Thus, the driver and the car form a hybrid assemblage (Featherstone, 2004). Likewise, Thrift mentions how the automobiles become more and more hybrid entities so that the separation between the human and machine becomes blurred (Thrift, 2004). Therefore, the driver – the owner – attributes different meanings to his or her car: he sees it as an object that symbolizes his social position and his 'being' within society. Thus, the society that we live in becomes a car society.

Moreover, the car offers different uses for the driver. It is 'a new form of communications platform with a complex set of possibilities' (Featherstone, 2004, p. 8). It can be used for accomplishing a set of activities within a specific time, and thus it becomes a platform for multi-tasking, which forms an important part in the discussions of women's perception of travel. The car allows women to accomplish their different roles in daily life: child care and active in work. Although the car is a tool that can increase women's freedom, it does so in the context of familial domesticity and becomes 'a tool to enact cultures of mothering' (Dowling, 2000).

Walking and cycling are at the heart of the research on mobility cultures. Leisure walking, walking to work or for exploring the city and different geographies, for producing literature and other artistic purposes are some examples of experiencing the time and space that we live in. As Middleton (2010) states, walking in the city creates unplanned interaction, unpredictable encounters between strangers. Thus, walking goes beyond being merely a transport mode and becomes a way of experiencing the socio-spatial arena. In today's cities, where the car dominates the roads, streets, squares and even the pavements, the most convenient places to walk are either green spaces or the historic town centres that restrict the movement of cars. Influenced by old towns, several local authorities or communities, many of which ironically are in the US, have developed design measures in order to make walking safer and more pleasurable. Websites promoting a more sociable and green community life rank the cities according to their walkability. For example, in the US the most walkable largest city is New York City followed by San Francisco and Boston (Walkscore, 2011).

Cycling and walking are often referred to as inexpensive ways of travelling and often attributed as travel modes for the poor. However, in Europe, especially in Scandinavian countries, cycling has an important place not only in leisure but also in urban trips. For example, in Denmark (and also in the Netherlands) cycling represents a crucial part of the transport culture (Jensen, 2007). Cycling produces similar experiences to walking as it 'provides us with an unbuffered range of sensory experiences of the monumental urbanity we have created' (Petersen, 2007, p. 37). Therefore, cycling is a way of understanding and experiencing the urban space as well as the countryside. Cycling creates encounters with other cyclists and pedestrians and it gives pleasure to the cyclist when cycling in a beautiful monumental city (Peteresen, 2007) such as cycling on the Golden Gate Bridge, which is a much more pleasurable way to discover the San Francisco Bay than being stuck in traffic for hours. Different accounts of cycling have appeared in the last couple of decades with research on the technology of cycling, bicycle activism, bicycle messengers, urban cycling, road racing, mountain biking and mediatization of cycling (Spinney, 2009). However, much of the emphasis is on the rise of environmentalism and the health benefits of cycling. Activist movements such as Critical Mass and groups such as the World Carfree Network and Sustrans (in the UK) are working towards a more sustainable future by drawing people's attention to other ways of travelling than by cars and planes and to better uses of urban space other than using it for parking (Critical Mass, 2011; Sustrans, 2011; World Carfree Network, 2011). Some of the actions being taken by increasingly growing numbers of environmentalists include activities such as legal road closures in order to celebrate urban life and decorating streets and using them for recreational purposes or just cycling on the roads with mass crowds, activities that can be regarded as protest activities.

Passenger mobilities are created within different modes of transport and offer different opportunities for interactions between people. As Bissell (2009b, p. 270) highlights in relation to travel by train, 'in the process of travel, we temporarily submit ourselves to become a part of a mobile collective'. Interaction and communication between 'mobile collectives' can take many forms and does not necessarily have to be vocal. Bissell draws our attention away from individual travel emotions to the context of interaction and its affect: '[a]ffect decentres the individual passenger from analysis, and instead prompts us to think about how different configurations of objects, technologies, and bodies come together to form different experiences of "being with" whilst on the move' (Bissell, 2009b, p. 270). Watts (2008) has examined interactions within the context of making space and time on a train trip across England by keeping a diary of her ten-hour train journey from Lancaster to St Ives. Her travelogue explores the art and craft of train travel, how passengers can also be understood as assemblages of property (clothes, money, music players, food and drink) that together shape their travel time use and make a destination. Similarly, Bissell (2009b) highlights that the sociality of the railway carriage is not formed only by individual bodies, although they are a part of it, but also constructed by music players, train tickets and seats.

## BUILDING A NEW MOBILITY CULTURE FOR A LOW CARBON SOCIETY

This chapter has discussed the main issues arising from contemporary mobilities research and the ways in which mobility cultures are performed. As stated in the introductory chapter to the book, the task of achieving a low carbon society is a difficult one to reconcile in relation to current mobility cultures. However, the future of mobility cultures may be formed differently since we can learn from past and current experiences. The sets of meanings, whether at the micro scale or the global, play an important role in the production of future mobilities (Cresswell, 2011). Therefore, lessons can be taken from the mobility cultures for achieving a low carbon society.

The main problem that is raised through the discussion is that the current mobility patterns are mainly built on social and economic inequalities. As we have seen in the examples, transport policies are dominated by 'male' and 'non-immigrant' (mainly middle-class) travel patterns. Policies ignore the travel patterns of women, the number of trips they have to do and the places they need to travel to. Likewise, the travel needs of lower-class citizens are not taken into consideration. Consequently, transport policies are developed around the dominance of car travel. However, as Hanson (2010, p. 16) suggests '[m]oving towards sustainable mobility will require improved understanding of how fully fleshed-out gender and fully fleshed-out mobility connect together differently in different contexts'. Therefore, the future of mobilities needs to take account of differences within society. Only then, can one of the important prerequisites of sustainable development, social sustainability, be fulfilled.

The rapid increase in urban population and the expansion of cities (see Chapter 5) create a growing demand for transport infrastructure, services and mobility levels, and, thus, sustaining low carbon mobilities becomes an ambitious task to accomplish. However, there is a growing interest and awareness in society about the crucial need to take action against environmental deprivation, including that resulting from transport. This awareness needs to be accounted for by local and central authorities and results in improved governance and making public participation available and continuous in helping to achieve sustainability goals.

An increasing number of academic studies highlight the need to understand how mobility cultures are formed not only in transportation research but also in anthropology, sociology, geography, urban studies and many other disciplines. It is quite clear that in order to understand mobility cultures and develop policies that will promote a shift towards a low carbon society, transportation research needs to collaborate with these other disciplines. The future of mobilities research will be based on understanding how mobility cultures are created through social norms, power, identity and equality and by different forms of transport, but also on the need for changing mobility habits as a consequence of issues such as climate change and the physical expansion of cities. As Hanson (2010, p. 7) points out, '[u]nderstanding everyday mobility is vital not only because it is fundamental to livelihoods, family life, community life – and to conceptions of gender . . . – but also because it is at the core of issues surrounding energy consumption, carbon emissions and settlement patterns, all of which are central to sustainability questions'.

## REFERENCES

Adey, P. (2010), *Mobility, Key Ideas in Geography*, London: Routledge.

Banister, D. (2008), 'The sustainable mobility paradigm', *Transport Policy*, **15**(2), 73–80.

Bissell, D. (2009a), 'Moving with others: the sociality of the railway journey', in P. Vannini (ed.), *The Cultures of Alternative Mobilities: Routes Less Travelled*, Aldershot: Ashgate, pp. 55–70.

Bissell, D. (2009b), 'Passenger mobilities: affective atmospheres and the sociality of public transport', *Environment and Planning D: Society and Space*, **28**, 270–89.

Church, A. and M. Frost (1999), *Transport and Social Exclusion*, Report for London Transport Planning.

Cresswell, T. (2010a), 'Towards a politics of mobility', *Environment and Planning D: Society and Space*, **28**, 17–31.

Cresswell, T. (2010b), 'Mobilities I: catching up', *Progress in Human Geography*, **35**(4), 550–8.

Cresswell, T. (2011), 'Mobilities II: still', *Progress in Human Geography*, online version, 1–9.

Cresswell, T. and P. Meriman (2011), *Geographies of Mobilities: Practices, Spaces, Subjects*, Aldershot: Ashgate.

Critical Mass (2011), *Critical Mass*, available at http://en.wikipedia.org/wiki/Critical_Mass (accessed 10 December 2011).

Dowling, R. (2000), 'Cultures of mothering and car use in suburban Sydney: a preliminary investigation', *Geoforum*, **31**, 345–53.

Featherstone, M. (2004), 'Automobilities: an introduction', *Theory Culture Society*, **21**, 1–24.

Hanson, S. (2010), 'Gender and mobility: new approaches for informing sustainability', *Gender, Place and Culture*, **17**(1), 5–23.

Jensen, O.B. (2007), 'Biking in the land of the car – clashes of mobility cultures in the USA', Paper presented at the 'Trafikdage' conference, Aalborg University, 27–28 August.

Jensen, O.B. (2009a), 'Foreword: Mobilities as culture', in P. Vannini (ed.), *The Cultures of Alternative Mobilities: Routes Less Travelled*, Aldershot: Ashgate, pp. xv–xix.

Jensen, O.B. (2009b), 'Flows of meaning, cultures of movements – urban mobility as meaningful everyday life practice', *Mobilities*, **4**(1), 139–58.
Le Guin, U. (1969), *The Left Hand of Darkness*, St Ives: Clays.
Lucas, K. (2004), *Running on Empty: Transport, Social Exclusion and Environmental Justice*, Bristol: The Policy Press.
Manderscheid, K. (2009), 'Unequal mobilities', in T. Ohmmacht, H. Maksim and M.M. Bergman (eds), *Mobilities and Equality*, Aldershot: Ashgate, pp. 27–50.
Massey, D. (1994), *Space, Place and Gender*, Minneapolis, MN: University of Minnesota Press.
Middleton, J. (2010), 'Sense and the city: exploring the embodied geographies of urban walking', *Social and Cultural Geography*, **11**(6), 575–96.
Petersen, J. (2007), 'Pedaling hope', *Magazine on Urbanism*, **6**, 36–9.
Rajé, F. (2004), *Transport Demand Management and Social Inclusion: The Need for Ethnic Perspectives*, Aldershot: Ashgate.
Sheller, M. and J. Urry (2006), 'The new mobilities paradigm', *Environment and Planning A*, **38**, 207–26.
Skeggs, B. (2004), *Class, Self, Culture*, London: Routledge.
Social Exclusion Unit (2003), *Making the Connections: Final Report on Transport and Social Exclusion*, UK: Social Exclusion Unit.
Spinney, J. (2009), 'Cycling the city: movement, meaning and method', *Geography Compass*, **3**(2), 817–35.
Sustrans (2011), *Sustrans*, available at http://www.sustrans.org.uk/ (accessed 10 December 2011).
Thrift, N. (2004), 'Driving in the city', *Theory, Culture and Society*, **21**(4/5), 41–59.
Urry, J. (2000), *Sociology Beyond Societies: Mobilities for the Twenty-first Century*, London: Routledge.
Urry, J. (2007a), 'Climate change, travel and complex futures', *British Journal of Sociology*, **59**(2), 261–79.
Urry, J. (2007b), *Mobilities*, Cambridge: Polity Press.
Uteng, T.P. (2009), 'Gender, ethnicity, and constrained mobility: insights into the resultant social exclusion', *Environment and Planning A*, **41**, 1055–71.
Uteng, T.P. and T. Cresswell (2008), *Gendered Mobilities*, Transport and Society: Ashgate.
Walkscore (2011), *Walkscore*, available at http://www.walkscore.com/rankings/ (accessed 10 December 2011).
Watts, L. (2008), 'The art and craft of train travel', *Social and Cultural Geography*, **9**(6), 711–26.
World Carfree Network (2011), *World Carfree Network*, available at http://world carfree.net/ (accessed 10 December 2011).

# 3. Accessibility, equity and transport

**Julia Markovich**

## INTRODUCTION

Reductions in the levels of greenhouse gas (GHG) emissions from the transport sector is an important overriding policy concern, but this must be achieved without compromising access to the services and amenities that are required for everyday life and contribute to social and economic inclusion (SSEE, 2010). This challenge is indicative of a broader policy conflict between sustainability and social equity objectives that has been observed in both the developed world (for example, Lucas, 2006) and in rapidly urbanizing and motoring countries (for example, Ahmed et al., 2008). While sustainability initiatives emphasize significant reductions in traffic levels and promote active travel and public transport, social equity initiatives recognize that the car is increasingly necessary for socially disadvantaged groups to fully participate in everyday life. The reductions in car use required to adequately decrease GHG emissions could thus drastically and adversely affect the life chances of many disadvantaged individuals and communities.

This chapter addresses the theme of equity and accessibility in transport, and comprises five sections. It aims to provide the reader with an in-depth analysis of the main issues associated with this theme and its intersection with wider issues, such as social inclusion. The second section briefly outlines some of the conceptual issues associated with equity and accessibility as they pertain to transport. This is followed by an overview of the importance of accessibility, particularly as it relates to the literature on transport and social exclusion. Noting the link between conceptual understandings and the implementation of transport policy (for example, Farrington, 2007), the fourth section provides an overview of some of the methodological challenges associated with measuring and evaluating accessibility. The final section summarizes the main research issues and challenges for the future of transport within the context of accessibility and equity.

## DEFINITIONS AND CONCEPTS

As many scholars have observed (see, for example, Geurs and van Wee, 2004; Litman, 2006; Farrington, 2007; Schwanen and De Jong, 2008; Geurs et al., 2009; Rosenbloom, 2009; TRB, 2011), there are a number of different conceptualizations for the terms 'equity' and 'accessibility'. Rosenbloom (2009), for example, recognizes more than 25 different definitions of equity in the literature on service delivery and infrastructure finance. Levinson (2002, p. 179) similarly notes the 'highly subjective' nature of this concept, one that often 'changes with the individual concerned'. Recent and comprehensive overviews of this term have further highlighted its multidimensional nature (see, for example, Barham and May, 2009; TRB, 2011).

While different criteria are used for defining equity, the notion of 'fairness' underlies this concept (TRB, 2011) and the two terms are often used interchangeably within the context of transport (see, for example, Sustainable Development Commission, 2011). A standard approach to equity in public finance has been described as one that considers 'the distribution of costs and benefits between groups and among individuals' (Schweitzer and Taylor, 2008, p. 798). Costs and benefits associated with a transport policy or initiative may be economic (such as a new tax or fee) or non-economic in nature (such as time lost or gained), and comprise short- and long-term impacts (TRB, 2011).

Rietveld (2003) identifies two main ways in which equity has a role in transport: (1) as an 'unintended side-effect' of policies designed to address other transport problems, such as environmental nuisance or congestion or (2) as an expressed aim or primary motivation for particular transport policies (such as the construction of infrastructure in disadvantaged regions or special facility provision for disabled people). Recognizing that there is no single definition of equity that 'can be considered the "right" one' (Rietveld, 2003, p. 595), Table 3.1 summarizes five common equity concepts, as well as their respective definitions and examples in practice (for comparable summaries see Litman, 2006; Rietveld, 2006).

Equity assessments in transport require the identification of impacts on different groups of people (such as by age, gender, income status, disability and other socio-demographic characteristics), and determining whether these impacts are fair or can be remedied (TRB, 2011). As Thomopoulos et al., (2009) observe, the emerging interest in the equity effects of transport projects is partly attributed to debates surrounding the introduction of road and congestion charging schemes (for a review of this literature see Levinson, 2010). Yet the distribution of these effects

*Table 3.1 Equity concepts*

| Type of equity | Simple definition | Transportation example |
|---|---|---|
| Benefits received | I get what I pay for. | People who use a facility the most pay the most. |
| Ability to pay (or income equity) | I pay more because I have more money. | A project is financed through a progressive tax that is disproportionately paid by higher income people. |
| Return to source (or geographic equity) | We get back what we put in. | Transit investment in each county is matched to that county's share of metropolitan tax revenues used for transit. |
| Costs imposed | I pay for the burden I impose on others. | Extra expense required to provide express bus service for suburb-to-city commuters is recovered through fares on this service. |
| Process (or participation) | I had a voice when the decision was made. | Public outreach regarding proposed new high occupancy toll lanes provides transparent information and seeks to involve all affected parties in public hearings and workshops. |

*Source:* TRB (2011, p. 41).

necessarily relates to numerous aspects of transport and is not limited to pricing and charging schemes.

Achieving fairness in accessibility for different groups is another important consideration in transportation equity (Sanchez et al., 2003). Within the context of passenger transport accessibility is defined as:

> (t)he extent to which land-use and transport systems enable (groups of) individuals to reach activities or destinations by means of a (combination of) transport mode(s). (Geurs and van Wee, 2004, p. 128)

More broadly, the term is used to refer to 'the freedom an individual has to participate in activities in the environment' (Casas, 2007, p. 464); the ability of individuals to 'get to key services at reasonable cost, in reasonable time and with reasonable ease' (SEU, 2003, p. 1). Recognizing the tendency within the literature for the terms 'accessibility' and 'access' to be used interchangeably, Geurs and van Wee (2004) distinguish between the former, which refers to a person's perspective, and the latter, referring to a location's perspective. There are a number of components to accessability that have been emphasized within the literature and Cass et al. (2005) provide what is arguably the most comprehensive overview of these at the present time (also see Geurs and van Wee, 2004 for a parallel typology). These are summarized in Table 3.2.

Concerns about inequity and accessibility in transport have centred on combating 'social exclusion' (Sanchez et al., 2003). While the origins of this term are associated with the work of French social scientists Lenoir and Lefebvre in the 1970s (Preston and Rajé, 2007), interest in the relationship between social exclusion and transport began to develop in the UK by the late 1990s and early 2000s (see Church et al., 2000; Lucas et al., 2001). As Cebollada (2009, p. 227) observes, social exclusion is a multidimensional concept and while there is no single accepted definition, it is viewed as involving:

> the lack or denial of resources, rights, goods and services, and the inability to participate in the normal relationships and activities, available to the majority of people in a society, whether in economic, social, cultural or political arenas. It affects both the quality of life of individuals and the equity and cohesion of society as a whole. (Levitas et al., 2007, p. 9)

While poverty is viewed as a contributing factor in social exclusion, Church et al. (2000) assert that the latter is a much broader concept that implies not only poverty, but the loss of ability to connect (literally and metaphorically) with employment, services and other facilities required to more fully participate in society. Seven transport-related categories of social exclusion are identified: physical; geographical; exclusion from facilities; economic exclusion; and time-, fear- and space-based exclusion (Church et al., 2000).

A number of different groups within society have been identified as being especially vulnerable to transport-related social exclusion. These include: ethnic minorities (for example, Rajé, 2004); women (PriyaUteng, 2009; Noack, 2011); lone parents (Páez et al., 2009); the elderly (Páez et al., 2009); young children and adults (Casas et al., 2009); people of low income (Páez et al., 2009, 2010); and individuals with physical disabilities (Casas, 2007).

## THE IMPORTANCE OF ACCESSIBILITY, TRANSPORT AND SOCIAL INCLUSION

As indicated in the previous section, the notion of access is essential to the work on transport and social exclusion, and a lack of access to important

*Table 3.2 Accessibility concepts*

| Accessibility dimension | Characteristics |
|---|---|
| Financial | • Financial resources are necessary for all forms of travel;<br>• Examples required for social inclusion: ownership/use of a car/taxi, availability of a 'point of contact' (e.g. telephone/mobile phone/email), ability to purchase intermittent long-distance travel, resources to stay over and meet with friends/family/colleagues while away;<br>• Recognized by local authorities via subsidies and concessionary passes. |
| Physical | • A function of design (e.g. cars, buildings, cities) and the capacities of users (especially for the physically impaired);<br>• Examples include: the inability to get into/drive a car, difficulties with walking certain distances or in unsafe areas; physical difficulties involved in entering different sites; limitations on the capacity to read timetabled information; physical constraints involved in carrying/moving heavy/large objects;<br>• Recognized in policy through the introduction of building regulations as well as those for transport vehicles, such as buses and trains to accommodate users with wheelchairs. |
| Organization | • Access to services/facilities depends on how transport is organized;<br>• Social inclusion with regards to the car is often dependent on social networks and being able to organize and negotiate lifts from others (e.g. family, friends, neighbours, colleagues);<br>• For those without access to a car, the organization of public transport raises such issues as: proximity to bus stops and train stations; direction of travel for buses; ability to reach a variety of destinations directly or indirectly; the quality of the travel experience; conditions of waiting and interchange locations and the transport service's frequency, reliability and punctuality. |
| Temporal | • Access is a matter of timing, time resources and time management;<br>• Temporal availability, such as the operating hours of public transport, incompatibility of transport services relative to the availability of residents to use a service, or services that have to be curtailed due to the time and frequency of services. |

*Table 3.2* (continued)

*Note:* While travel time is arguably the most basic and important dimension of accessibility, the interpretation of this dimension as cited above has recently been challenged by Banister (2011, p. 950), who articulates a 'more flexible interpretation of time constraints' within the context of travel. He argues that the commodification of time has led to a preoccupation with travel time savings over all other aspects. Yet, as Metz (2008) documents using National Travel Survey (NTS) data for the UK, time saved is instead used for additional travel over longer distances, enabling people to gain access to a greater range of destinations. Successive increases in distances travelled observed over the last 30 years have ceased in this last decade, leading Metz (2010) to conclude that a saturation for the demand for travel can be expected. Recognizing the limitations of the preoccupation with travel time savings, Banister further argues that other elements, such as the quality of time and the benefits of slower travel over shorter distances, should receive greater attention. As journeys over shorter distances are more amenable to the use of walking, cycling and public transport, Banister (2011) argues that an emphasis on slower travel and shorter journeys would challenge the dominant car culture and have important environmental and social benefits, thus resolving the policy tension between sustainability and equity objectives outlined earlier in this chapter.

*Source:* Adapted from Cass et al. (2005, pp. 549–51).

amenities and services associated with everyday life, or 'accessibility deprivation' (Neutens et al., 2011), is a central feature of transport-related social exclusion. Accordingly, and as Cass et al. (2005) confirm, the concept of accessibility has become: 'increasingly influential in current thinking about the causes and consequences of social exclusion'. It has been described as 'an index of exclusion' (Casas, 2007), and is valued for its role in enhancing an understanding of how people are able to participate more fully in the activities associated with everyday life (Páez et al., 2010).

According to Farrington (2007), accessibility became accepted as a transport policy goal within the UK Government largely as a result of the Social Exclusion Unit's (2003) seminal report *Making the Connections*, which broadly framed transport as a social policy issue and more specifically established the linkages between lack of access to transport and the exclusion of members of society from important life chances. The role of transport in enabling people to participate in various activities, such as employment and learning opportunities, and to make use of local services, such as healthcare, food shops and recreational and leisure facilities, received particular attention within this report (SEU, 2003).

Transport problems in developing countries have also been linked to extreme inequality, low incomes, and rapid population growth and urbanization, with the demand for urban transportation infrastructure and services far exceeding supply (Pucher et al., 2005). The significant increase in the use of private cars, particularly within large cities, has created a range of negative transport externalities in developing countries (Ahmed et al., 2008). As the urban poor of these countries cannot generally afford access to motorized transport, they are increasingly disadvantaged as a result of their relative immobility and inaccessibility (Pucher et al., 2005). As Pucher et al. (2005, pp. 185–6) describe, the majority of developing countries suffer from the following access issues: the absence of an effective transport infrastructure and other important services, resulting from rapid and unplanned suburban development; poor quality road networks that are often unpaved and not maintained; high levels of traffic congestion with motorized and non-motorized vehicles sharing roads at highly variable speeds; rapidly rising use and ownership of cars and motorcycles; inadequate provision for buses and non-motorized modes of transport; poor traffic control/management; high levels of traffic fatalities, especially for motorcyclists and pedestrians; poor public transport; and very high levels of environmental pollution and noise derived from transport.

While recognizing that transport alone cannot eliminate poverty, the World Bank (2006) views it as central to its mission of poverty reduction. To improve mobility and accessibility within developing countries, which for lower income groups appears to be in decline, a number of public policies have been identified. These include: rights of ways for pedestrians and cyclists; improved traffic management and public transport services; privatization of bus services; road design that accommodates multiple modes (especially buses, cyclists and pedestrians); improvements to motor vehicle technology and fuels; transit-oriented development; and increased funding for transport systems more generally (Pucher et al., 2005). There is also an important economic dimension to access, particularly for developing countries. The World Bank (2006), for example, recognizes the integral link between access to transport and a country's economic performance; transport infrastructure plays a key role in enabling poorer members of society to gain access to job opportunities and in stimulating economic growth.[1]

Improvements to transport and access, such as those cited above, can thus make a meaningful contribution to the social inclusion of disadvantaged individuals and communities. And while social inclusion is related to many issues beyond transport, improved access to transport is seen to help individuals overcome issues associated with social exclusion by enabling access to employment opportunities, health and other activities, thus helping people to feel more included (Mackett et al., 2008). As Lucas et al. (2001, p. v) observe: 'transport often acts as an enabler of greater social inclusion, in that increased mobility can stimulate a virtuous cycle of enhanced accessibility and social and economic engagement'.

It is important to recognize, however, that two main assumptions implicit in the conceptualization of accessibility (as in Table 3.2) and its relationship to social exclusion are increasingly being challenged by transport scholars. The first is the emphasis on physical mobility as the outcome associated with enhancing access and as the means of reducing social exclusion. Whether through improved financial resources, physical design, organization or timing of services, the primary objective associated with improving access as outlined in Table 3.2 is to increase the level of physical mobility of individuals relative to the services and amenities required to support social inclusion. This emphasis on physical mobility negates the potential contribution that other forms such as virtual mobility might make in reducing social exclusion (see Kenyon et al., 2002). While an understanding of this contribution is admittedly still in its infancy (Kenyon, 2006), the interest in virtual mobility suggests opportunities for further research and a potential way through the sustainability and social equity policy tension described earlier.

Second, and related to this, is the presumption that accessibility is always positive, that 'more is always seen as better' (Ross, 2000, p. 14), and that the key to achieving more accessibility is to increase (physical) mobility. This presumption overlooks the problem of too much mobility, one that has profound social and spatial consequences (for example, Adams, 2001) and can further contribute to social exclusion (for example, Preston and Rajé, 2007). As Adams (2001) illustrates, the problems associated with hypermobility have been overlooked relative to its advantages. Widespread use of the car has already created, and will continue to create, suburban sprawl, thus reinforcing the need for additional car-based travel (Adams, 2001). This has the added effect of subordinating:

> other 'public' mobilities of walking, cycling, travelling by rail, and so on; and it reorganises how people negotiate the opportunities for, and constraints upon, work, family life, leisure and pleasure. (Sheller and Urry, 2003, p. 173)

Too much mobility can also contribute to the social exclusion of individuals and communities 'through environmental degradation, adverse public health impacts, high accident rates, declining public transport, changes in land use and community severance' (Preston and Rajé, 2007, p. 154). The issue of forced car ownership for low-income households in peripheral communities (for example, Dodson and Sipe, 2007) in order to access the essential services and amenities of everyday life can also been viewed as one way in which too much mobility contributes to transport disadvantage, transport poverty and further exclusion. Recognizing the negative consequences for both sustainability and social inclusion objectives, a re-evaluation of the ways in which accessibility and mobility are defined may be useful in addressing the aforementioned policy tension. Couclelis (2000), for example, introduces the notion of 'sustainable accessibility' and the related goal of 'reasonable mobility' as preconditions for sustainable transportation, thus engaging with these issues. Banister (2011, p. 958) further calls for a 'reorientation of priorities in transport planning', emphasizing the benefits of shorter distances, slower travel and more local activities, thereby reducing the need for private transport.

## MEASUREMENT OF ACCESSIBILITY

Given the importance of accessibility in better understanding transport-related social exclusion (and inclusion), an accepted approach to measuring this concept is critical to furthering research and policy in this area (Church and Marston, 2003). There remain, however, a number of issues with respect to measuring and evaluating accessibility. These include: the effectiveness of conventional versus more innovative methods; the need to tailor accessibility measures to the mobility requirements of particular disadvantaged populations; the development of measures that go beyond considering the spatial organization of land uses and include other dimensions, such as operating hours; and the development of accessibility measures for non-motorized modes, such as walking and cycling, and the role of modes such as taxis.

A number of scholars have elaborated on the variety of methods that exist for measuring accessibility, the limitations associated with more conventional methods and the opportunities that exist to develop or adopt new approaches (for example, Geurs and van Wee, 2004; Preston and Rajé, 2007; Doi et al., 2008; Schwanen and De Jong, 2008; Casas et al., 2009; Páez et al., 2010; and others). Schwanen and De Jong (2008) distinguish between two broad approaches to measuring accessibility: 'conventional accessibility' (CA) measures, such as gravity-based and cumulative-opportunity measures,[2] and 'space-time accessibility' (STA) measures,[3] derived from Hägerstrand's (1970) time-geography approach (what Geurs and van Wee refer to as 'person-based measures'). As Páez et al. (2010) observe, accessibility research has increasingly involved more refined indicators, and moved beyond defining accessibility at an aggregate level (as in conventional, location-based measures) towards an individual accessibility framework informed by the time-geography approach.

Conventional and STA approaches each have their own limitations in practice. Conventional measures that evaluate accessibility at an aggregate level are valued for the generalizability of their findings and their ability to assess wider trends (for example, Páez et al., 2010), yet they are unable to capture more intricate and individual differences in accessibility in the way that STA approaches can (Schwanen and de Jong, 2008; Casas et al., 2009; Neutens et al., 2011). By contrast, STA measures that evaluate individual mobility have proved useful in exploratory analyses and provide rich and detailed data, yet they are not easily applied to large population samples and thus lack the generalizability associated with conventional methods (Páez et al., 2010; Neutens et al., 2011).

As Geurs and van Wee (2004, p. 129) observe, the type of accessibility measure employed is necessarily related to the accessibility dimension being studied. These scholars argue that the temporal dimension of accessibility, for example, is best evaluated using person-based (STA) measures; other approaches to measuring accessibility do not generally consider this dimension (ibid. p. 130). The acceptance of various measurement approaches is thus directly related to the ways in which the concept of accessibility is defined.

Related to this first issue is that there is some evidence within the literature to suggest that accessibility measures should necessarily be tailored to the mobility requirements of particular disadvantaged populations, such as in the case of physically disabled individuals (see Church and Marston, 2003; Casas, 2007), children (for example, Casas et al., 2009) and the elderly (Mackett et al., 2008). As an alternative to absolute measures of accessibility, for example, Church and Marston (2003) introduce the concept of 'relative accessibility' for physically disabled individuals, a measure that relates differential access between groups of users (for example, a person in a wheelchair relative to an ambulatory person) and can more generally assist with understanding what physical impairments and constraints need to be overcome in travel and throughout the built environment.

Similarly, Casas et al.'s. (2009) comparison of three different methods used to identify transport-related social exclusion in children (a conventional technique, based on a modification of a traditional deprivation index; a cumulative accessibility measure, which reflects an individual's activity space; and space-time prisms, describing how activities are coordinated in space based on time and other constraints) reveals that each method differentially identified excluded groups within their sample. These methods were used to evaluate travel diary data for a sample of 674 children between 5 and 18 years in the counties of Erie and Niagara in New York, US. The results of this evaluation revealed that the traditional deprivation index approach more often identified children as being 'not excluded' when the cumulative accessibility and space-time prism approaches confirmed that they are excluded. In addition, the space-time approach was 'methodologically more sound' than the other approaches, and better able to consider such issues as differences in age and changes in travel mode over time.

Unique approaches to assessing the accessibility of the elderly have also been advocated. Mackett et al. (2008), for example, have pioneered a new software tool called A Methodology for Enhancing Life by Increasing Accessibility (AMELIA), an interface to a geographic information system (GIS) that helps assess how transport policies can advance social inclusion and help planners determine which accessibility options are the most cost-effective. Their pilot application of this software in the assessment of four transport policy options in the city of St Albans, UK reveals the adaptability of this methodology, and the tool developed, when engaging with the various mobility needs of people 65 years of age and older (such as wheelchair users) and in improving access to the city centre and town hall. The introduction of benches was found to be the most cost-effective approach for improving accessibility in this study, and the overall findings suggest significant potential for applying this methodology and tool in other contexts.

An additional and more recent body of research suggests that accessibility measures have been predominantly focused on the spatial organization of land uses and thus overlook other dimensions of accessibility, such as temporal considerations. Studies by Delafontaine et al. (2011) and Neutens et al. (in press), for example, are part of a recent but growing body of literature that examines the role of public service operating hours and accessibility. This research suggests that there are particular opportunities for rescheduling the operating hours of many public services in order to improve the accessibility of individuals, and more generally highlights a role for temporal planning that is as significant as land use or transport planning in advancing social inclusion.

Lastly, as Iacono et al. (2009) observe, the emphasis on accessibility measures within transportation planning is largely placed on motorized modes. Non-motorized modes, such as walking and cycling, have received less attention within the literature, though this is changing (see, for instance, Pucher and Buehler, 2008; Heinen et al., 2010). As these scholars assert, data quality issues remain one of the main sources of difficulty in evaluating the accessibility of non-motorized modes. Lack of non-motorized travel behaviour data; limited high-resolution land use data; inadequate zonal structure and travel network data; and the use of arbitrary impedance functions (which reflect the influence of travel time, money and other costs on an individual's willingness to travel greater distances) for walking and cycling have all been cited as difficulties in developing accessibility measures for these modes (Iacono et al., 2009), thus highlighting an important area for further research. Mulley's (2010) recent research on the potential role of national taxi-based services in improving accessibility in rural areas and on the periphery of urban areas similarly suggests that there is an opportunity to consider extensions to the current public transport mix as a means of reducing social exclusion. The role of taxis and non-motorized modes may prove most useful in addressing the mobility and accessibility issues associated with an ageing population, as driving cessation often becomes a reality for many older people and public transport is generally seen as ill-suited to this age group, particularly for those living outside city centres (Rosenbloom, 2010). An emphasis on non-motorized modes also addresses concerns about too much (private) mobility addressed in the previous section.

## CONCLUDING REMARKS

As car dependency has become a central feature of everyday life, particularly for socially and economically disadvantaged members of society, the task of achieving GHG emission reduction targets from the transport sector necessarily requires engaging with a range of equity and accessibility issues. The future of mobility will thus rest on resolving the policy tension between sustainability initiatives, which emphasize a reduction in motorized traffic levels, and equity perspectives, which recognize the value of car-based travel in reducing social exclusion. This final section highlights some of the key elements that will shape the future of mobility with respect to equity and accessibility.

As illustrated in this chapter, equity, accessibility and social inclusion/exclusion are complex and multidimensional in nature and thus a number of conceptual issues have emerged. Recognizing the linkages between conceptual understandings, methodological challenges and the implementation of transport policy, a key challenge for transport research will necessarily involve refining the definitions of these terms to ensure greater clarity and consistency within academic and policy literature.

A number of methodological challenges have also been identified in this chapter. These are partly derived from the conceptual issues described previously, and yet also relate to broader debates about the role of conventional versus more innovative methods, and the development of approaches that are tailored to both the accessibility dimension being evaluated and the particular needs of excluded populations. Earlier calls for an accepted set of approaches to measuring accessibility are thus still relevant today and for the future (Church and Marston, 2003).

In addition, a number of research gaps and opportunities for further scholarly work in the area of equity, accessibility and transport have been identified. These include consideration of a greater range of accessibility dimensions (such as temporal accessibility and an emphasis on virtual mobility) and evaluation of the full range of modes (including non-motorized modes and an expanded notion of public transport services to include opportunities for taxis).

It is becoming increasingly apparent that equity and accessibility implications associated with transport are important considerations in achieving a low carbon society. As illustrated in this chapter, key debates and issues necessarily cut across a range of disciplines, not only in transport but in urban studies, geography, sociology, economics and other fields. Looking forward, two broad challenges remain for transport scholars and policy-makers. The first is enhanced political and financial support for policies that address equity and accessibility concerns reducing GHG emissions from the transport sector. The need for increased political support is an issue that has been raised in the previous decade (for example, Pucher et al., 2005), and is most certainly an issue today in light of widespread austerity measures and cutbacks. The second is the need for an interdisciplinary framework to help extend current thinking and ensure dialogue amongst disciplines. It is this collaborative framework that will ultimately guide the future of transport research, such that a low carbon (and equitable) society can be achieved.

## NOTES

1. Recent research on the construction of selected large-scale transport infrastructure projects in developing countries contests the presumption of economic benefits. Writing on the case of the Delhi Metro, for example, Siemiatycki (2006) cites an inherent 'paradox of progress' associated with the development, including: land appropriations and community displacements, protests and petitions from displaced residents, neighbourhood revitalization schemes that favour elite interests, and further divides between the city's rich and poor inhabitants. Richardson and Jensen's (2008) research on the Bangkok Skytrain similarly reveals that this transport project caters principally to Thai middle-class shoppers, the Thai business community and tourists, as lower income residents and migrant workers are unable to afford tickets.
2. Schwanen and de Jong (2008, p. 559) describe three main elements to gravity-based and cumulative opportunity measures: (1) a reference location, used to determine access to all other locations; (2) a set of opportunities or destinations (e.g. shops, jobs, services); and (3) an impedance function that captures the effect of separation on destination attractiveness. Whereas gravity-based measures employ a distance-decay function to discount the attractiveness of destinations further away, cumulative opportunity measures use the impedance function to exclude opportunities 'beyond a specified threshold value' (Schwanen and de Jong, 2008, p. 559).
3. STA measures derived from Hägerstrand's (1970) time-geography approach are based on the assumption that individuals follow a path through space and time, and that this path can be visualized (on a two-dimensional plane using space on the horizontal

axis and time on the vertical axis, and by using GIS software). They rely on time diary data provided by individuals as to their daily routine, indicating destination, mode and purpose of travel.

## REFERENCES


Adams, J. (2001), 'The social consequences of hypermobility', RSA Lecture 21, November, available at http://john-adams.co.uk/wp-content/uploads/2006/hypermobilityforRSA.pdf (accessed 15 January 2012).

Ahmed, Q.I., H. Lu and S. Ye (2008), 'Urban transportation and equity: a case study of Beijing and Karachi', *Transportation Research Part A,* **42**, 125–39.

Banister, D. (2011), 'The trilogy of distance, time and speed', *Journal of Transport Geography,* **19**, 950–9.

Barham, P. and A. May (2009), *CURACAO Deliverable D2: State of the Art Review (FINAL)*, Leeds: Institute for Transport Studies, University of Leeds, Chapter 10.

Casas, I. (2007), 'Social exclusion and the disabled: an accessibility approach', *The Professional Geographer*, **59**(4), 463–77.

Casas, I., M.W. Horner and J. Weber (2009), 'A comparison of three methods for identifying transport-based exclusion: a case study of children's access to urban opportunities in Erie and Niagara counties, New York', *International Journal of Sustainable Transportation*, **3**, 227–45.

Cass, N., E. Shove and J. Urry (2005), 'Social exclusion, mobility and access', *Sociological Review,* **53**(3), 539–55.

Cebollada, A. (2009), 'Mobility and labour market exclusion in the Barcelona Metropolitan Region', *Journal of Transport Geography*, **17**, 226–33.

Church, A., M. Frost and K. Sullivan (2000), 'Transport and social exclusion in London', *Transport Policy*, **7**(3), 195–205.

Church, R.L. and J.R. Marston (2003), 'Measuring accessibility for people with a disability', *Geographical Analysis*, **35**(1), 83–96.

Couclelis, H. (2000), 'From sustainable transportation to sustainable accessibility: can we avoid a new tragedy of the commons?', in D. Janelle and D. Hodge (eds), *Information, Place, and Cyberspace: Issues in Accessibility*, Berlin: Springer-Verlag, pp. 342–56.

Delafontaine, M., T. Neutens, T. Schwanen and N. Van de Weghe (2011), 'The impact of opening hours on the equity of individual space-time accessibility', *Computers, Environment and Urban Systems*, **35**(4), 276–88.

Dodson, J. and N. Sipe (2007), 'Oil vulnerability in the Australian city: assessing socioeconomic risks from higher urban fuel prices', *Urban Studies*, **44**(1), 37–62.

Doi, K., M. Kii and H. Nakanishi (2008), 'An integrated evaluation method of accessibility, quality of life, and social interaction', *Environment and Planning B: Planning and Design*, **35**(6), 1098–116.

Farrington, J. (2007), 'The new narrative of accessibility: its potential contribution to discourses in (transport) geography', *Journal of Transport Geography*, **15**(5), 319–30.

Geurs, K.T. and B. van Wee (2004), 'Accessibility evaluation of land-use and transport strategies: review and research directions', *Journal of Transport Geography*, **12**, 127–40.

Geurs, K.T., W. Boon and B. van Wee (2009), 'Social impacts of transport: literature review and the state of the practice of transport appraisal in the Netherlands and the United Kingdom', *Transport Reviews*, **29**(1), 69–90.
Hägerstrand, T. (1970), 'What about people in regional science?', *Papers of the Regional Science Association*, **24**, 7–21.
Heinen, E., B. van Wee and K. Maat (2010), 'Commuting by bicycle: an overview of the literature', *Transport Reviews*, **30**(1), 59–96.
Iacono, M., K. Krizek and A. El-Geneidy (2009), 'Measuring non-motorized accessibility: issues, alternatives, and execution', *Journal of Transport Geography*, **18**(1), 133–40.
Kenyon, S. (2006), 'Reshaping patterns of mobility and exclusion? The impact of virtual mobility upon accessibility, mobility and social exclusion', in M. Sheller and J. Urry (eds), *Mobile Technologies of the City*, New York: Routledge, pp. 102–20.
Kenyon, S., G. Lyons and J. Rafferty (2002), 'Transport and social exclusion: investigating the possibility of promoting inclusion through virtual mobility', *Journal of Transport Geography*, **10**(3), 207–19.
Levinson, D. (2002), 'Identifying winners and losers in transportation', *Transportation Research Record*, **1812**, 179–85.
Levinson, D. (2010), 'Equity effects of road pricing: a review', *Transport Reviews*, **30**(1), 33–57.
Levitas, R., C. Pantazis, E. Fahmy, D. Gordon, E. Lloyd and D. Patsios (2007), *The Multi-dimensional Analysis of Social Exclusion*, Bristol: Department of Sociology and School for Social Policy, Townsend Centre for the International Study of Poverty and Bristol Institute for Public Affairs, University of Bristol.
Litman, T. (2006), *Evaluating Transportation Equity Guidance For Incorporating Distributional Impacts in Transportation Planning*, Victoria: Victoria Transport Policy Institute.
Lucas, K. (2006), 'Providing transport for social inclusion within a framework for environmental justice in the UK', *Transportation Research Part A*, **40**, 801–9.
Lucas, K., T. Grosvenor and R. Simpson (2001), *Transport, the Environment and Social Exclusion*, York: Joseph Rowntree Foundation and York Publishing.
Mackett, R., K. Achuthan and H. Titheridge (2008), 'AMELIA: a tool to make transport policies more socially inclusive', *Transport Policy,* **15**, 372–8.
Metz, D. (2008), 'The myth of travel time saving', *Transport Reviews*, **28**(3), 321–36.
Metz, D. (2010), 'Saturation of demand for daily travel', *Transport Reviews*, **30**(5), 659–74.
Mulley, C. (2010), 'Promoting social inclusion in a deregulated environment: extending accessibility using collective taxi-based services', *Research in Transportation Economics*, **29**, 296–303.
Neutens, T., T. Schwanen and F.Witlox (2011), 'The prism of everyday life: towards a new research agenda for time geography', *Transport Reviews*, **31**(1), 25–47.
Neutens, T., M. Delafontaine, T. Schwanen and N. Van de Weghe (in press), 'The relationship between opening hours and accessibility of public service delivery', *Journal of Transportation Geography*, doi:10.1016/j.jtrangeo.2011.03.004.
Noack, E. (2011), 'Are rural women mobility deprived? – a case study from Scotland', *SociologiaRuralis*, **51**(1), 79–97.
Páez, A., R.G. Mercado, S. Farber, C. Morency and M. Roorda (2009), *Mobility*

*and Social Exclusion in Canadian Communities: An Empirical Investigation of Opportunity Access and Deprivation from the Perspective of Vulnerable Groups*, Toronto, Canada: Policy Research Directorate, Strategic Policy and Research.
Páez, A., R.G. Mercado, S. Farber, C. Morency and M. Roorda (2010), 'Relative accessibility deprivation indicators for urban settings: definitions and application to food deserts in Montreal', *Urban Studies*, **47**(7), 1415–38.
Preston, J. and F. Rajé (2007), 'Accessibility, mobility and transport-related social exclusion', *Journal of Transport Geography*, **15**, 151–60.
PriyaUteng, T. (2009), 'Gender, ethnicity and constrained mobility: insights into the resultant exclusion', *Environment and Planning A*, **41**, 1055–71.
Pucher, J. and R. Buehler (2008), 'Making cycling irresistible: lessons from the Netherlands, Denmark and Germany', *Transport Reviews*, **28**(4), 495–528.
Pucher, J., N. Korattyswaropam, N. Mittal and N. Ittyerah (2005), 'Urban transport crisis in India', *Transport Policy*, **12**, 185–98.
Rajé, F. (2004), *Transport Demand Management and Social Inclusion: The Need for Ethnic Perspectives*, Aldershot: Ashgate.
Richardson, T. and O.B. Jensen (2008), 'How mobility systems produce inequality: making mobile subject types on the Bangkok Sky Train', *Built Environment*, **34**(2), 218–31.
Rietveld, P. (2003), 'Winners and losers in transport policy: on efficiency, equity, and compensation', in D.A. Hensher and K.J. Button (eds), *Handbook of Transport and the Environment*, Amsterdam: Elsevier, pp. 585–602.
Rietveld, P. (2006), 'Pricing in transport; a multimodal perspective. An introduction', *European Transport*, **32**, 1–4.
Rosenbloom, S. (2009), 'The equity implications of financing the nation's surface transportation system', *TR News*, **261**, 3–9.
Rosenbloom, S. (2010), 'How adult children in the UK and the US view the driving cessation of their parents: is a policy window opening?', *Journal of Transport Geography*, **18**, 634–41.
Ross, W. (2000), 'Mobility & accessibility: the yin & yang of planning', *World Transport Policy and Practice*, **6**(2), 13–19.
Sanchez, T.W., R. Stolz and J.S. Ma (2003), *Moving to Equity: Addressing Inequitable Effects of Transportation Policies on Minorities*, Cambridge, MA: The Civil Rights Project at Harvard University.
Schwanen, T. and T. De Jong (2008), 'Exploring the juggling of responsibilities with space-time accessibility analysis', *Urban Geography*, **29**(6), 556–80.
Schweitzer, L. and B.D. Taylor (2008), 'Just pricing: the distributional effects of congestion pricing and sales taxes', *Transportation*, **35**(6), 797–812.
Sheller, M. and J. Urry (2003), 'The city and the car', in A. Root (ed.), *Delivering Sustainable Transport: A Social Science Perspective*, London: Routledge, pp. 171–90.
Siemiatycki, M. (2006), 'Message in a metro: building urban rail infrastructure and image in Delhi, India', *International Journal of Urban and Regional Research*, **30**(2), 277–92.
SSEE (Smith School of Enterprise and the Environment) (2010), *Future of Mobility Roadmap: Ways to Reduce Emissions while Keeping Mobile*, Oxford: Smith School of Enterprise and the Environment.
SEU (Social Exclusion Unit) (2003), *Making the Connections: Final Report on Transport and Social Exclusion*, London: Office of the Deputy Prime Minister.

Sustainable Development Commission (2011), *Fairness in a Car-dependent Society*, London: Sustainable Development Commission.

Thomopoulos, N., S. Grant-Muller and M.R. Tight (2009), 'Incorporating equity considerations in transport infrastructure evaluation: current practice and a proposed methodology', *Evaluation and Program Planning*, **32**, 351–9.

TRB (Transportation Research Board) (2011), 'Committee on equity implications of evolving transportation finance mechanisms', *Equity of Evolving Transportation Finance Mechanisms*, TRB Special Report 303.

World Bank (2006), *Social Analysis in Transport Projects: Guidelines for Incorporating Social Dimensions into Bank-supported Projects*, Washington, DC: Social Development Department, The World Bank.

# 4. Governance, policy and mobility futures

**James Macmillen**

## INTRODUCTION

Without a comprehensive understanding of mobility governance and attendant policies, well-intentioned efforts to engender transformational change in contemporary mobility regimes are prone to failure.[1] Indeed, considerations of governance straddle the two fundamental pillars supporting intellectual efforts in the field of transport research: the need to understand contemporary mobility regimes and the need to expediently intervene in such regimes in order to realize political ends. Importantly, it must be recognised that modes of mobility governance cannot be divorced from the broader (mobility) cultures within which they are embedded (see Chapter 2), and vice versa. Through a continual process of structuration (cf. Giddens, 1984), mobility governance is at once shaped by structural, cultural dynamics and, in turn, shapes contemporary mobility practices through its own profound agency. Public infrastructure projects – such as the development of the US highway network – are obvious material manifestations of such agency, while historic legislative actions – such as the 1944 Chicago Convention on International Civil Aviation – illustrate the longevity of certain governmental practices and the pronounced degree of systemic path dependency they can generate.

At the outset, it is helpful to define the key concepts drawn upon and discussed in this chapter. Essentially, 'governance' is to be understood as broader than the modus operandi of the state, or 'government'. For while 'government signals a concern with the formal institutions and structures of the state, the concept of governance . . . draws attention to the ways in which governmental and nongovernmental organizations work together, and to the ways in which political power is distributed, both internal and external to the state' (Goodwin, 2009, p. 593). Conversely, 'policy' is here understood as referring to 'a series of intentionally coherent decisions or activities taken or carried out by different public – and sometimes private – actors whose resources, institutional links and interests vary, with a view to resolving in a targeted manner a problem that is politically defined as collective in nature' (Knoepfel et al., 2007, p. 24). This broader interpretation of policy is adopted to distinguish it from specific policy 'interventions', those courses of action formulated and implemented in order to reach particular objectives.

In responding to the threat of anthropogenic climate change, the need to analyse the nature of governance in the transport domain is abundantly clear. While historical evidence shows that major sociotechnical transitions involve myriad cultural, technological and institutional components, there is broad agreement that achieving a genuine transition to a global low carbon mobility regime necessitates effective, efficient and coordinated policy action at all spatial scales and across geographical jurisdictions, from supra-national frameworks – such as the European Union (EU) emissions trading scheme – to the local authority provision of low carbon infrastructure. Regrettably, however, prevailing conceptualizations of governance in the transport field tend to be relatively naïve and under-theorized, partly reflecting the discipline's long-standing affinity with the engineering sciences and neoclassical economics (Banister et al., 2011). This has profound implications for how policy interventions are formulated, implemented and appraised, and has resulted in something of a professional consensus in which technological solutions, compact urban design and pricing instruments are framed as panacean (Banister et al., 2011). Indeed, even where the centrality of governance is recognized, emergent from the literature are often sets of generic 'policy recommendations' implicitly directed to 'government' in the abstract, rarely integrated into a discussion of specific governance architectures.

This chapter proceeds through four sections. First, an overarching conceptual discussion is offered that seeks to clarify the nature of contemporary mobility governance and corresponding policy interventions. Second, a number of salient research challenges pertaining to mobility governance and policy are critically discussed. Third, drawing lightly on the notion of sociotechnical regimes, a future outlook on mobility governance is presented. Finally, some concluding remarks are offered.

## GOVERNANCE AND POLICY: A SYSTEMIC PERSPECTIVE

As implied above, the particular disciplinary heritage of transport studies has had a profound effect upon the manner in which mobility governance has been conceptualized. Specifically, there is a dominant tendency to view transport policy-making in a relatively narrow, technical sense, grounded in instrumental rationality and conceived of as linear in character. First, so the assumed model goes, transport-related 'problems', such as congestion or environmental pollution, are generally framed as self-evident and technical in character; policy objectives are then formulated in relation to these problems – directly or indirectly relating to the economy, environment, accessibility and so on; third, policy interventions are formulated in order to meet these objectives in the most effective, efficient and acceptable means possible; fourth, these interventions are implemented, and their success evaluated. Ostensibly, this does seem intuitive. Indeed, despite heavy critique, instrumental rationality is a fundamentally important element in policy-making; few, for example, would sensibly question the value of designing policy interventions that are effective and efficient (although it must be recognized that these terms are inherently political and open to diverse interpretation). Moreover, although the linear, 'stages model' of the policy process bears little relation to the flux and 'irrationalities' of real-world policy-making, its value as a heuristic device for structuring both critical discussion and 'process advocacy' contributions should not be underestimated (Hill, 2009).

However, this traditional, 'policy-based' perspective is far from optimal. Simply put, it fails to adequately engage with the nuanced character and complex geographies of contemporary mobility governance. Importantly, this is not to say that tacit proponents of the linear model completely fail to acknowledge the presence of such complexity per se; many do so, especially in relation to issues of policy interventions' legitimacy and economic feasibility. Rather, there is a pronounced tendency to invoke a crude dualism between 'politics', on the one hand, and 'policy', on the other. This fuels an attendant belief in a division of labour where 'political' complexity is treated as somehow exogenous to the responsibilities and concerns of those, in the transport field, tasked with supporting and informing the design of actual policy interventions.

A governance perspective, however, offers a holistic, systemic perspective on mobility politics and policy and avoids such compartmentalization. Fundamentally, processes of decision-making and policy-making in transport are no longer simply interpreted as straightforward exercises in technical planning, but rather as processes 'of exchange, transaction and bargaining between different institutions and policy actors' (Ney, 2009, p. 27; see also Banister et al., 2011). According to this view, the 'sites' of mobility governance are located within a policy subsystem characterized both by inordinate technical complexity and entrenched value-laden disagreement, symptomatic of 'wicked' (Rittel and Webber, 1973) or 'messy' (Ney, 2009) policy problems. Significantly, the antagonism and intractability present in such arenas stems from the power-laden interplay of various 'advocacy coalitions' – alliances of functionally heterogeneous actors that mobilize around particular 'frame'-orientated core beliefs in order to exert a concerted influence on the policy process.[2] Undermining the assumptions of the traditional conceptualizations of transport policy, the actions and inactions of these coalitions are not restricted to the 'political' realm of agenda-setting, but simultaneously permeate all 'stages' of the policy process (see Sabatier and Weible, 2007).

Although inevitably stylized, Ney (2009) identifies three major advocacy coalitions operating in the transport/mobility policy domain, which he dubs 'efficient mobility', 'sustainable mobility' and 'balanced mobility', respectively. The first, Ney argues, is characterized by a position that holds mobility to be wholly subservient to the demands of economic growth. Correspondingly, the scope for transport policy is limited to the provision of sufficient infrastructure such that continued future demand for mobility may be accommodated. The second, in marked contrast, views contemporary mobility regimes as environmentally exploitative and socially unjust, symptomatic of overarching socioeconomic order that it also augments and entrenches. The scope for policy here is much greater, with redistributive and regulatory interventions prioritized. Finally, the third coalition attempts a degree of dialectical mediation between the other two, recognizing both the benefits and costs of mobility. The corresponding task for policy is to manage mobility in a systematic and coherent manner so as to achieve a satisfactory degree of 'equilibrium' between economic, societal and environmental interests.

As stressed in the introduction, the diverse makeup of these advocacy coalitions is testament to the fact that governance is a much more embracing concept than 'government'. Myriad state and non-state actors, such as politicians, bureaucrats, scientific experts, interest groups, the media, commercial interests, non-governmental organizations and the citizenry, are enmeshed in a complex array of institutionally fragmented, formal and informal relationships, characteristic of what Ansell (2000) terms the 'networked polity'. In contrast to traditional, hierarchical modes of government – where policy domains were characterized by 'many to one' associations between non-state actors and the state – the relationships linking actors across the contemporary networked polity in transport governance are more heterarchical, or 'many to many', in nature, serving to embed the state in, and ensure its reliance upon, both civil society and market systems (Smith et al., 2005; Ney, 2009). Moreover, as Marsden and Rye (2010) note, the issue of scale is highly significant (cf. the third section). Transport governance is profoundly 'multi-level', involving decision-making at a range of spatial scales, from the supra-national to the local, all the while involving the participation of non-state actors (Bache and Flinders, 2004).

Regardless of how fragmented and contested contemporary transport governance may be, however, various ends do come to be officially pursued by the state. Indeed, the whole enterprise of public policy is concerned, at least ostensibly, with formulating and implementing interventions that seek to expediently influence certain societal practices towards desired ends. Much of the academic literature concerned with the policy process conceives of such ends as 'objectives'. Yet, arguably the connotations of this term present a slightly misleading picture. Desired ends may well be clear-cut, comprehensively articulated and/or measurable 'objectives' – framed, for example, in relation to the legally binding reductions in UK carbon dioxide ($CO_2$) emissions – but these are perhaps the exception to the rule. On the contrary, reflecting the profound degree of realpolitik in the transport domain, policy interventions are regularly enacted in the service of all manner of vague, implicit and politically 'safe' aspirations, such as the abstract desire for time- and cost-efficient transport networks.

Mentioned above, policy interventions are strategic compositions of one or more policy measures or instruments, defined by Bemelmans-Videc et al. (1998, p. 3) as 'techniques by which governmental authorities wield their power in attempting to ensure support and effect social change'. Within the transport domain we can point to three major families of routinely deployed policy measures (Vedung, 1998). The first, are 'regulatory measures', and refer to rules, directives and standards that oblige certain actors to behave in a certain manner. Second are 'economic measures' pertaining to the allocation of material resources, either in cash or in kind, and acting as incentives or disincentives. Thus, while this would include outright fiscal taxes and subsidies, it also covers non-monetary material resources, such as transport infrastructure provision. Third are 'informative measures', involving attempts to persuade actors to modify their behaviour on the basis of moral appeals and reasoned argument. This rough typology will be drawn upon further in the fourth section.

Importantly, in the quest to bring about genuine low carbon mobility, it is clear that the scope of analytical efforts must encompass more than a narrow concern for the present governance of mobility per se. Rather, we are interested in purposeful, systemic regime change, which, by definition, requires myriad responses – conscious or otherwise – from a plethora of actors and institutions across spatial scales and geographical contexts (Smith et al., 2005). Literature on sociotechnical transitions is particularly informative here. We can think of a regime as a 'series of complex, nested real world phenomena, embodying natural and artificial physical elements, as well as social, economic, cultural and cognitive attributes' (Smith et al., 2005, p. 1493; see also Rip and Kemp, 1998). It is tempting to treat the entirety of global mobility as one such regime. However, it is more appropriate to consider a range of contiguous regimes, differentiating on the basis of some core characteristic(s), such as where modes of transport are qualitatively distinct. Hence we might speak of an aviation-based 'aeromobility' regime as conceptually distinct from a car-based 'automobility' regime, although in the low carbon mobility context they share many characteristics. Indeed, we might well characterize the prevailing land-based mobility paradigm in the advanced capitalist economies as 'hydrocarbon-dependent automobility'. 'Hydrocarbon-dependent' as its continuity is almost entirely contingent on the discovery, extraction, refinement, freightage and combustion of petrochemicals; and 'automobility', following Urry (2004), as it is largely predicated on quasi-autonomous travel practices afforded by the private automobile. As Urry (2004) also notes, the only 'certainty' surrounding such complex, non-linear autopoietic regimes is that there is no certainty with respect to their future evolution. This simultaneously presents both a challenge and an opportunity to governance and policy; change is incredibly hard to manage, but certain to occur.

Highlighting the importance of sociotechnical regimes in this manner also allows us to consider the interdependencies that exist between mobility governance and the international governance practices surrounding anthropogenic climate change and the shift to a low carbon future. Critically, within this overarching governance architecture, there exists a set of major challenges that mobility governance – particularly in relation to low carbon transport – directly inherits. Drawing on Carter's (2007) discussion of environmental politics, for example, we can see how efforts to curb carbon emissions from transport are hampered by standard collective action problems (such as free-riding), as well as the salience of transboundary conditions and issues of spatial and temporal variability in both the source and impact of emissions.

## SALIENT ISSUES AND RESEARCH CHALLENGES

As with any brief overview, the following discussion on high-level research issues and challenges facing the field of sustainable mobility governance is necessarily partial. At the expense of a fully comprehensive account, it instead concentrates on two core issues that are particularly prevalent in the literature: (1) addressing deficits of policy principle and scope and (2) understanding the geographies of governance.

**Ontological Deficits in Principles and Scope**

Essentially, this challenge relates to the assumptions guiding policy formulation and the depth and breadth of associated conceptual frameworks. The first area of interest pertains to the relatively narrow range of principles and metrics that support decision-making in the transport domain, augmented by the failure of policy to appreciate the significance of path dependencies and systemic lock-in to incumbent mobility regimes. Such ontological deficits are particularly evident in mainstream understandings of travel behaviour. As Cresswell (2006) and many others are at pains to point out, the phenomenon of mobility is much more intricate and multifaceted than the simple 'transport' of people and freight from point A to point B (see Chapter 2). However, much of the thinking underlying transport governance and policy remains wedded to a strongly utilitarian perspective that fails to adequately reflect this complexity.

The standard, functionalist treatment of time in policy formulation and appraisal is particularly illustrative here. Reflecting the capitalist imperative for time-space compression, concern for travel time has been at the heart of transport studies since its inception. The central assumption of this concern is that time spent travelling is inherently wasted time, as it cannot be used productively for economic gain, despite some evidence to the contrary (Holley et al., 2008; Jain and Lyons, 2008). Consequently, much transport policy has sought to minimize travel time through the deployment of interventions that facilitate increased speed (Banister, 2011). However, time 'saved' through faster travel is by no means allocated directly into more 'productive' activities. Instead, as many analyses have shown, travellers typically exhibit a relatively constant 'travel time budget' of somewhere between 60 and 90 minutes per day, and correspondingly use 'time saved' to travel greater distances (see, for example, Metz, 2008a), a development largely at odds with the goal of low carbon mobility (Metz, 2008a; Banister, 2011; see also Marchetti, 1994).

The implications for governance are stark. First, and most immediately, conceptual understandings of travel time, speed and distance need to be significantly rethought. Travel time must be fundamentally reinterpreted in a manner sensitive to the embodied and relational nature of mobility, and the fact that experiences and attitudes to travel time are not universal, but rather, as cultural constructs, differ profoundly across space and time (Banister, 2011). Second, there are concomitant and pressing issues relating to the justification and appraisal of transport schemes. Presently, estimated travel time savings account for approximately 80 per cent of the benefits resulting from traffic schemes (Metz, 2008b). Adopting a more nuanced conceptualization of travel time hence demands a more reflexive and creative approach to evaluation. Is it acceptable, as happens at present in the UK, for small savings in the travel time of individuals with high market values of time to be equated with large travel time savings for those with a low market value of time (see Buchan, 2009)? Should travel time savings be treated equally across demographic groups or modes of transport (cf. equity concerns in Chapter 3)?

Similar ontological challenges for research are encountered in relation to what might be termed policy interventions' scope. First, in relation to 'lateral' scope, the relatively narrow conceptual basis of transport policy unfortunately manifests itself in a perspective that tends only to recognize policy interventions' immediate impacts in relation to pre-specified principles, aims, indicators and contexts. Clearly, given the complexities and interdependencies of contemporary mobility regimes, a much greater effort is required to be able to comprehensively consider the impact of interventions in their totality. Of particular significance is the issue of unintended effects (Gudmundsson et al., 2010), where well-intentioned policy interventions aimed at one particular problem can inadvertently affect others. Naturally, these may directly arise in relation to interventions' immediate impacts – where, for example, policy measures perversely influence their specific objectives – but they may also arise in relation to other transport objectives, or objectives in other policy domains entirely. Catalytic converters, for example, were introduced to reduce emissions of nitrogen oxides, a cause of acid rain. However, this measure had the unintended consequence of reducing engine efficiency and hence resulted in greater $CO_2$ emissions for a given distance travelled (Carter, 2007). Interestingly in terms of governance, unintended effects also arise through particular approaches to policy formulation and implementation in addition to the mechanisms of interventions per se. Examining practices of project planning and appraisal, Flyvbjerg (2009) illustrates how theoretically optimal processes of competitive tendering for large public infrastructure projects can actually result in major cost overruns and benefit shortfalls.

Second, challenges also arise in relation to interventions' 'temporal' scope, insofar as sound policy development requires comprehensive and accurate *ex ante* and *ex post* appraisals of interventions' effectiveness and efficiency. Unsurprisingly, the vast majority of *ex ante* appraisal is concerned with interventions' immediate effectiveness and associated financial costs. However, also reflecting limited 'lateral scope', such appraisal rarely exhibits a concern for issues of process and the influence of complicating factors in the policy process that fall somewhat outside the traditional purview of transport analysis. The issue of legitimacy barriers to coercive, repressive policy measures neatly illustrates this. For example, although its scheme-level strategy was technically strong, the proposed Manchester congestion charging scheme in the UK was overwhelmingly rejected by the citizenry on the basis of a referendum. A challenge for research, therefore, is to arrive at more holistic and politically savvy modes of *ex ante* appraisal (Givoni et al., 2010). Conversely, *ex post* scheme appraisal is rarely conducted at all, and when it is, it can reveal how inaccurate and over-optimistic *ex ante* forecasts can be (Flyvbjerg et al., 2003).

**Governance Geographies**

As Anable and Shaw (2007, p.451) observe, 'the notion of scale, as in many other environmental governance debates, is vital to considerations of how to reduce transport emissions'. In one sense, there is a problematic disjuncture between the inherently fluid and mobile nature of transport practices and the relatively rigid, formalized institutional structures that hope to manage such practices (Marsden and Rye, 2010). In another sense, critiques of rigid scalar conceptualizations of governance are vital in order to understand and manage present mobility regimes. However, they should not be so strong as to override the pressing need for strategic modes of governance to be developed at distinct scales (Anable and Shaw, 2007), and nor should they jeopardize an understanding of the relationships between scalar systems. Focusing on the UK and the potential for demand management interventions, Marsden and Rye (2010) posit a distinct lack of knowledge and clarity concerning respective responsibilities at various hierarchical levels of governance. As a result, they argue, the nation's overall strategic capacity for achieving transformative change is significantly diminished. The task of coherent governance, however, is certainly not aided by the fluidity of mobility practices. It is a significant challenge in itself to satisfactorily attribute carbon emissions to specific jurisdictional domains, let alone formulate comprehensive strategies to mitigate them. Where, for example, ought the carbon emissions from a car journey be registered? To the motorist's home region? To the destination? To areas travelled through en route (Marsden and Rye, 2010)? This same principle is active on larger scales too; because issues of cross-boundary emissions are so contentious and complex, it was one of the reasons why international maritime and aviation were not included in the Kyoto Protocol (Marsden and Rye, 2010).

For Vedung (1998) it is only through ensuring compatibility between supra-national, national, regional, sub-regional and local levels of governance that 'vertical packaging' of complementary policy measures at different jurisdictional scales can occur. Even where scalar responsibilities are clear, however, governance efforts can be 'usurped' by decisions made up and down the hierarchy. For example, voluntary agreements made between the European Commission and major automobile manufacturers concerning vehicular emissions standards consequently render it difficult for member states to enact further policy interventions at the national scale (Anable and Shaw, 2007). Surpra-national governance is, of course, vital, given the globalized nature of mobility and the uneven spatial and temporal impacts of anthropogenic climate change. Local governance is evidently crucially important also. However, it is not just these extremes of 'local' governance and 'global' governance that should be of interest in the transport policy domain, but rather the entire governance continuum, identifying and exploring linkages and the potential for knowledge transfer across scales (Anable and Shaw, 2007; see also Bache and Flinders, 2004). It is in this spirit that a re-emphasis on the city-regional scale has been staunchly advocated. As regions often represent travel to work areas, and exhibit a range of social and economic functions, they offer a strong potential as sites for strategic policy intervention. However, for city-regions to function as viable 'convergence spaces' (Routledge, 2003) for meaningful policy action, a concomitant devolution of policy formulation and implementation powers from the national scale is required (Anable and Shaw, 2007). In practice, the success of devolutionary action is variable and remains highly contingent upon the actions of key political actors (MacKinnon et al., 2010).

## FUTURE OUTLOOK

In considering the future of transport governance and policy through the lens of sociotechnical regime change, this section draws upon Smith et al.'s (2005) two concepts of: (1) selection pressures – forces that come to bear on mobility regimes; and (2) adaptive capacities for regime transformation – a function of resource availability and the potential for coordinated action in light of these pressures. Interpreted as sociotechnical regimes – configurations of objects, organizations, institutions, practices and networks – contemporary mobility practices appear to exhibit a significant degree of stability. Yet this impression of stability is less an enduring construct, and rather the product of shifting selection pressures coming to bear on the regime and the extent to which endogenous and exogenous resources are able to adapt to such pressures (Smith et al., 2005).

Importantly, given that there are innumerable selection pressures on regimes at any one time, it is the articulation of particular pressures that is of significance, for this is what largely determines their transformative capacity (Smith et al., 2005). Therefore, rather than reviewing articulated selection pressures that further the project of low carbon mobility – such as the scientific consensus on anthropogenic climate change – the focus here concerns perhaps the most significant selection pressure working across contemporary mobility regimes that acts against the project of low carbon mobility: simply, the apparently insatiable demand for mobility in the modern world, and the absence of political commitment to move towards low carbon transport systems associated with it.[3]

Throughout the advanced capitalist economies mobility has long been directly associated with economic growth, freedom, wealth, status and quality of life, with the result that hydrocarbon-dependent automobility is the established mobility regime for land-based passenger transport (Urry, 2004). Although North America, Australasia and most of Europe epitomize this state of affairs, rates of urbanization and automobile ownership are currently rising rapidly in the emerging economies of Asia and South America (Kenworthy, 2006). In relation to low carbon mobility, this growth in demand has significant consequences. Indeed, such is the growth in mobility in the EU27 in recent years that the attendant growth in carbon emissions has dramatically undermined the environmental benefits resulting from the emergence of clean technologies and increased vehicular efficiencies. As Nykvist and Whitmarsh (2008, p. 1374) observe, 'the benefit of technical measures to reduce vehicle emissions . . . has often been outstripped by the increase in vehicle numbers, engine size, travel frequency and trip length'.

The theoretical rationale behind the principle of demand management and associated interventions, therefore, is widely acknowledged. As Goodwin et al. (1991) argued in their highly cited report to the Rees Jefferys Road Fund, titled *Transport: The New Realism*, the unquestioning logic of 'predict and provide' in transport policy that dominated the latter half of the twentieth century is a grossly insufficient basis for policy over the medium to long term (see Owens, 1995). However, in seeking to address the form and intensity of mobility demand through existing techniques of governance, policy-makers have faced many obstacles. We may return to Vedung's (1998) rough typology of policy measures outlined in the second section, recalling that policy measures may be regulatory, economic or informative in character. Recognizing that single measures deployed in isolation are rarely effective, policy interventions are typically comprised of a range of measures – although the degree to which such combinations are to be considered 'optimal' is often debatable (Givoni et al., 2010). Importantly, in relation to demand management, both stimulative and repressive variants exist for each measure type. Road pricing schemes, for example, are generally considered to represent the theoretically optimal means of managing demand; they enable the internalization of motoring's external costs, and can support the flexible management of congested systems through cordon placement and timeframe selection. In design, they are inherently repressive. Specifically, they simultaneously incorporate repressive variants of regulatory measures – insofar as motorists must adhere to various proscriptions and prohibitions associated with such schemes – and repressive variants of economic measures, insofar as road pricing functions as an in-cash disincentive, acting to discourage motorists from driving in particular locations at particular times. The essential point relates to the fact that while coercive, repressive policy measures are theoretically highly effective, they have a marked tendency to invoke significant hostility when they are proposed. Without a threshold degree of legitimacy, such interventions are incredibly difficult to implement (Mahendra, 2008; Gullberg and Isaksson, 2010). Indeed, as the case of the failed Manchester congestion charging scheme profoundly illustrates, even where such interventions are technically well developed and integrated, they are highly susceptible to political attack (Sherriff, 2009).

Of course, road pricing is not a panacea to the problems of insatiable demand. Mobility networks are highly adaptive systems, and liable to produce unintended effects when narrowly acted upon, particularly where the consumer or end-user is the sole focus for policy intervention at the expense of other significant agents (for example, vehicle manufacturers). More significantly, road pricing is not primarily designed to mitigate carbon emissions; traffic may be redistributed onto other areas of the road network, speeds may increase that lead to exponentially higher emissions and decongestion may lead to a reduction in the overall cost of motoring, thereby stimulating additional demand (Anable and Shaw, 2007). Indeed, there are a range of additional measures commonly invoked in this context, including mixed, high density land use planning, encouraging the uptake of public transport and active travel, and improving the public realm through urban design and planning (Kenworthy, 2006). Informative, or so-called 'soft' measures, too, such as personalized travel planning and marketing can prove to be effective means of enacting voluntary change (Bamberga et al., 2011).

It is also important to consider the extent to which the adaptive capacities of existing structures of governance are able to satisfactorily support coordinated and strategic policy responses in light of selection pressures. Adaptive capacities relate to the resources of knowledge, materials and capabilities required for systemic change and the extent to which these responses can be coordinated by actors. As Banister et al. (2011) observe, transport governance is plagued by the legacy of Fordist convictions and associated divisions of labour that render the formulation and implementation of coordinated, holistic interventions very difficult. In many countries, for example, land use planning is controlled at the sub-regional scale, whereas strategic transport projects are publicly and privately financed at the national level, and managed at the sub-national level by ancillary authorities. The complexities of such 'vertical' power relations are further compounded by obfuscated 'horizontal' power relations, concerning the division of closely aligned responsibilities across multiple agencies at the same jurisdictional scale (Banister et al. 2011).

Drawing, albeit critically, on Unruh (2000), Banister et al. (2011) further highlight how policy-makers exhibit a distinct tendency to favour so-called 'end-of-pipe' policy measures in the first instance. Such measures shy away from engendering systemic change and instead aim to treat systems' associated symptoms. Examples include promoting technical advances in fuel and engine efficiency (Anable and Shaw, 2007). Only where such measures are insufficiently effective are 'continuity' measures considered – those measures that involve moderate intra-system change without disrupting the overall system structure, such as the encouragement of modest modal shift (Anable and Shaw, 2007) or facilitating the emergence of alternative propulsion technologies. Rarely are 'discontinuity' measures considered or implemented – those which involve radical overhauls of an existing regime, such as the development of international high-speed rail connections (Banister et al., 2011). In many ways, this sequential progression makes sense. As Vedung (1998) argues, cautious 'chronological' policy packaging from benign to coercive policy measures reflects both pragmatic, bureaucratic concerns as well as those that stem from liberal political philosophy. However, in relation to adaptive capacity, and leaving aside the difficulties inherent in 'process advocacy', following a linear progression from technological measures is by no means ideal. As Anable and Shaw (2007) make clear, for a genuine transition to sustainable mobility to occur, policy interventions must address innumerable shortcomings in the present mobility regimes, a vast number of which cannot be solved through technological fixes.

## CONCLUSIONS

This chapter has sought to clarify the nature of mobility governance and policy and ground it in the conceptual literature on sociotechnical regime change. It has stressed the need to acknowledge the complicated realities present in the governance of 'messy' or 'wicked' contemporary policy problems and to avoid treating them as exogenous to the concerns of transport academics, policy-makers and practitioners. Major challenges for research relate to the ontological basis of transport policy and the scope of its associated tools and techniques. In addition, awareness of the geographies of governance is crucially important if systemic change is to be genuinely realized across spatial scales and jurisdictional domains. At the heart of present mobility governance a tension exists between the globalized nature of mobility, and its attendant commercial interests, and the compartmentalized, siloed governance regimes that seek to influence them. This is inescapable to a degree, but requires concerted effort to avoid outright policy failure. With respect to selection pressures and adaptive capacities, the insatiable demand for mobility is presently offsetting the improvements realized through technological efficiency. A range of measures are potentially valuable in managing demand, but those that are most potentially effective are also those that are politically toxic. This illustrates the need to complement relatively narrow modes of scheme appraisal to take into account the likely critical reception of theoretically optimal policy interventions 'on the ground'. Additionally, mobility governance suffers from a legacy of compartmentalization and fragmentation, both vertically across spatial scales and horizontally, across jurisdictional domains.

Importantly, the prospects for global transformational change towards low carbon mobility cannot be considered and appraised in isolation. As this chapter has stressed, mobility governance is inherently conditioned by broader sociocultural, economic and demographic developments, many of which pose significant challenges to the realization of a low carbon economy as a whole. Indeed, while environmental advocacy coalitions have gained increasing degrees of political influence in recent decades, and while the sustainable development discourse is increasingly commonplace at a range of governmental scales, the dominant paradigm in environmental policy appears to be one of 'ecological modernization', where a focus on technological progress is prioritized as a means of simultaneously achieving economic growth while minimizing environmental harms (see also Chapter 13). The inherent conservatism of this approach to transport governance naturally shies away from significant, and hence politically contentious, change. Maintenance of the status quo is further augmented by challenges of communication, interpretation and trust between governance actors, commercial interests and the citizenry. As stressed, transport policy is far from a value-free technocratic exercise. Rather, particular policies and objectives will explicitly or implicitly reflect the interests of particular social groups and institutions. Such is the complexity of the issues involved that wider perceptions of particular policies may well differ significantly from their official function.

To conclude, although theoretically appealing, ecological modernization effectively weds mobility governance to the broader project of consumer capitalism and, as such, the reduction in transport emissions to be gained from technological developments are proving to be offset by the growth in demand for increasingly mobile lifestyles.

## NOTES

1. The term 'mobility regime' refers to the constellation of actors, technologies, institutions and infrastructures that collectively act to perpetuate distinct mobility practices over time and across space. In preference to narrower, more conventional, terminology in transport studies – such as 'travel patterns' – its usage in this chapter is intended to frame the complex, systemic and interdependent nature of such practices as constitutive of particular sociotechnical regimes (see Smith et al., 2005 and Chapter 14).
2. 'Cognitive frames' may be defined as perspectives 'from which an amorphous, ill-defined, problematic situation can be made sense of and acted upon . . . [affording] a way of selecting, organising, interpreting, and making sense of a complex reality to provide guideposts for knowing, analysing, persuading and acting' (Rein and Schön, 1993, p. 146).
3. For example, while the European Commission's recent Transport White Paper includes ambitious targets to 'halve the use of "conventionally-fuelled" cars in urban transport by 2030; phase them out in cities by 2050; [and] achieve essentially CO2-free city logistics in major urban centres by 2030' (European Commission, 2011, p. 9), it also explicitly states that 'curbing mobility is not an option' (p. 5).

## REFERENCES

Anable, J. and J. Shaw (2007), 'Priorities, policies and (time)scales: the delivery of emissions reductions in the UK transport sector', *Area*, **39**(4), 443–57.

Ansell, C. (2000), 'The networked polity: regional development in western Europe', *Governance*, **13**(2), 279–91.

Bache, I. and M. Flinders (eds) (2004), *Multi-level Governance*, Oxford: Oxford University Press.

Bamberga, S., S. Fujiib, M. Frimanc and T. Gärling (2011), 'Behaviour theory and soft transport policy measures', *Transport Policy*, **18**(1), 228–35.

Banister, D. (2011), 'The trilogy of distance, speed and time', *Journal of Transport Geography*, **19**(4), 950–9.

Banister, D., K. Anderton, D. Bonilla, M. Givoni and T. Schwanen (2011), 'Transport and the environment', *Annual Review of Environment and Resources*, **36**, 247–70.

Bemelmans-Videc, M.L., R.C. Rist and E. Vedung (eds) (1998), *Carrots, Sticks and Sermons: Policy Instruments and their Evaluation*, New Brunswick, NJ: Transaction.

Buchan, K. (2009), *Investing in Transport: Making the Change*, London: Green Alliance.

Carter, N. (2007), *The Politics of the Environment: Ideas, Activism, Policy*, Cambridge: Cambridge University Press.

Cresswell, T. (2006), *On the Move: Mobility in the Modern Western World*, Abingdon: Routledge.

European Commission (2011), *Roadmap to a Single European Transport Area – Towards a Competitive and Resource Efficient Transport System*, Transport White Paper, Brussels: European Commission.
Flyvbjerg, B. (2009), 'Survival of the unfittest: why the worst infrastructure gets built – and what we can do about it', *Oxford Review of Economic Policy*, **25**(3), 344–67.
Flyvbjerg, B., M.K. Skamris Holm and S.L. Buhl (2003), 'How common and how large are cost overruns in transport infrastructure projects?', *Transport Reviews*, **23**(1), 71–88.
Giddens, A. (1984), *The Constitution of Society*, Berkeley, CA: University of California Press.
Givoni, M., J. Macmillen and D. Banister (2010), 'From individual policies to policy packaging', Paper presented at the European Transport Conference, Glasgow, October.
Goodwin, P., S. Hallett, F. Kenny and G. Stokes (1991), *Transport: The New Realism. Report to the Rees Jeffreys Road Fund*, Transport Studies Unit, University of Oxford.
Goodwin, M. (2009), 'Governance', in R. Kitchin and N. Thrift (eds), *International Encyclopaedia of Human Geography*, Oxford: Elsevier, pp. 593–9.
Gudmundsson, H., C.H. Sørensen, J. Åkerman et al. (2010), 'Limits to and unintended effects of transport policies', Paper presented at the European Transport Conference, Glasgow, October.
Gullberg, A. and K. Isaksson (2010), *Congestion Taxes in City Traffic: Lessons Learnt from the Stockholm Trial*, Lund: Nordic Academic Press.
Hill, M. (2009), *The Public Policy Process*, Harlow: Longman.
Holley, D., J. Jain and G. Lyons (2008), 'Understanding business travel time use and its place in the working day', *Time and Society*, **17**(1), 27–46.
Jain, J. and G. Lyons (2008), 'The gift of travel time', *Journal of Transport Geography*, **16**, 81–9.
Kenworthy, J.R. (2006), 'The eco-city: ten key transport and planning dimensions for sustainable city development', *Environment and Urbanization*, **18**(1), 67–85.
Knoepfel, P., C. Larrue and M. Hill (2007), *Public Policy Analyses*, Bristol: The Policy Press.
MacKinnon, D., J. Shaw and I. Docherty (2010), 'Devolution as process: institutional structures, state personnel and transport policy in the United Kingdom', *Space and Polity*, **14**(3), 271–87.
Mahendra, A. (2008), 'Vehicle restrictions in 4 Latin American cities: is congestion pricing possible?', *Transport Reviews*, **28**(1), 105–33.
Marchetti, C. (1994), 'Anthropological invariants in travel behaviour', *Technological Forecasting and Social Change*, **47**(1), 75–88.
Marsden, G. and T. Rye (2010), 'The governance of transport and climate change', *Journal of Transport Geography*, **18**(6), 669–78.
Metz, D. (2008a), 'The myth of travel time savings', *Transport Reviews*, **28**(3), 321–36.
Metz, D. (2008b), *The Limits to Travel*, London: Earthscan.
Ney, S. (2009), *Resolving Messy Policy Problems: Handling Conflict in Environmental, Transport, Health and Ageing Policy*, London: Earthscan.
Nykvist, B. and L. Whitmarsh (2008), 'A multi-level analysis of sustainable mobility transitions: niche development in the UK and Sweden', *Technological Forecasting and Social Change*, **75**, 1373–87.

Owens, S. (1995), 'From "predict and provide" to "predict and prevent"? Pricing and planning in transport policy', *Transport Policy*, **2**(1), 43–9.
Rein, M. and D. Schön (1993), 'Reframing policy discourse', in F. Fischer and J. Forester (eds), *The Argumentative Turn in Policy Analysis and Planning*, Durham, NC: Duke University Press, pp. 145–66.
Rip, A. and R. Kemp (1998), 'Technological change', in S. Rayner and E. Malone (eds), *Human Choices and Climate Change*, vol. 2, Columbus, OH: Battelle, pp. 327–99.
Rittel, H. and M. Weber (1973), 'Dilemmas in a general theory of planning policy sciences', *Policy Sciences*, **4**(2), 155–69.
Routledge, P. (2003), 'Convergence space: process geographies of grassroots globalization networks', *Transactions of the Institute of British Geographers*, **28**, 333–49.
Sabatier, P. and C. Weible (2007), 'The advocacy coalition framework: innovations and clarifications', in P. Sabatier (ed.), *Theories of the Policy Process*, Boulder, CO: Westview, pp. 189–220.
Sherriff, G. (2009), 'Voting on good transport? Just sustainability after the Greater Manchester TIF Referendum', Paper presented at the AESOP Young Academics Conference, Vienna.
Smith, A., A. Stirling and F. Berkhout (2005), 'The governance of sustainable socio-technical transitions', *Research Policy*, **34**, 1491–510.
Unruh, G.C. (2000), 'Understanding carbon lock-in', *Energy Policy*, **28**, 817–30.
Urry, J. (2004), 'The "system" of automobility', *Theory, Culture and Society*, **21**(4–5), 25–39.
Vedung, E. (1998), 'Policy instruments: typologies and theories', in M. Bemelmans-Videc, R. Rist and E. Vedung (eds), *Carrots, Sticks and Sermons: Policy Instruments and their Evaluation*, New Brunswick, NJ: Transaction, pp. 21–58.

# 5. Urbanization and future mobility

**Robin Hickman**

## THE URBAN PHENOMENON AND TRAVEL

For the first time in human history, more than half of the world's population lives in cities, a remarkable rise in urbanization, relative to just 33 per cent in 1960 and 39 per cent in 1980 (World Bank, 2010). The city as the centre and 'milieu' of innovation is described, historically, as developing from the Athenian Marketplace of Plato and Socrates, to the Renaissance of Florence, the Industrial Revolution of Manchester and Birmingham, the Production of Mobility in Detroit, the Freeway of Los Angeles, to the Silicon Age of San Francisco and Bangalore, the Social Democracy of Stockholm and the Financial Capitalism of London and New York (Hall, 1988, 1998; Glaeser, 2011). In contemporary terms, there are now ten metacities (with over 20 million residents), including Seoul, Shanghai, Delhi, Mumbai and São Paulo, with over 36 million residents in the Tokyo Metropolitan Area. The megacity (over ten million residents) is home to 10 per cent of the world's population. Table 5.1 illustrates how the rate of urbanization varies globally, with Asia in particular at lower levels than the West, and population densities being very low in North and Latin America. In environmental terms, carbon dioxide ($CO_2$) emissions and energy use per capita (aggregate) are very high in North America and Europe. The ultimate 'triumph of the city' is in enabling collaboration and the flow of ideas, at times offering the potential for humanity to 'shine most brightly' (Glaeser, 2011). Yet, of course, there are severe problems experienced in many cities; with poor health, deprivation, low incomes, inadequate housing and employment; and high rates of emissions and resource consumption.

The urban phenomenon also has major implications for transport. The city, in theory, allows mass public transport systems to be built, but this is reliant on the shape and form of the urban area. Europe, with relatively high levels of urbanization and high densities, is associated with relatively low levels of fuel consumption per capita in transport, despite high vehicle ownership rates. North America, in particular, and also Latin America,

*Table 5.1 Regional differences in urbanization and energy use*

| Indicator | EU | North America | East Asia & Pacific | South Asia | Latin America & Caribbean | Sub- Saharan Africa | World |
|---|---|---|---|---|---|---|---|
| GDP per capita (current US$) | 36615 | 46786 | 6295 | 993 | 7845 | 1245 | 9161 |
| Urban population (% of total) | 73.69 | 81.58 | 47.23 | 29.50 | 78.56 | 36.45 | 50.01 |
| Population density (people per sq. km of land area) | 119.3 | 18.5 | 89.3 | 321.9 | 28.6 | 34.4 | 51.6 |
| $CO_2$ emissions (metric tons per capita) | 7.83 | 17.78 | 4.95 | 2.91 | 1.28 | 0.85 | – |
| Energy use (kg of oil equivalent per capita) | 3510 | 7532 | 1689 | 498 | 1312 | 710 | 1840 |
| Motor vehicles (per 1000 people) | 549 | 789 | 104 | 16 | 175 | – | – |
| Road sector gasoline fuel consumption per capita (kt of oil equivalent) | 0.19 | 1.12 | 0.08 | 0.01 | 0.14 | 0.03 | 0.14 |

*Source:* World Bank (2010).

show high levels of fuel consumption per capita. Asia and Africa have low vehicle ownership rates (as of now) and consequently low levels of fuel consumption per capita (Table 5.1).

One of the major fields of research within urban and transport planning, extending over the last 30 years, has been in examining the linkages between urban form, layout and travel. There has been a recent emphasis in the field on a particular angle: that of achieving greater sustainability in travel, often measured in terms of reducing transport $CO_2$ emissions (Ewing et al., 2008; Hickman et al. 2009). A very important policy driver, one that gives a much greater impetus to this research area, is climate change. Ambitious targets have been adopted internationally to reduce $CO_2$ emissions. For example, the UK is committed to an 80 per cent reduction in cross-sectoral $CO_2$ emissions by 2050 on a 1990 base (through the Climate Change Act, 2008) and various carbon budgets have been set over time, that is, the allowable 'spend' of carbon at the national level (Committee on Climate Change, 2010). The major difficulty, however, is in meeting these targets.

This chapter examines the role of urban structure in enabling future sustainable mobility. It considers the early and emerging literature, the main research issues and challenges that might help shape transport in the future. The discussion recognizes that there are many other socio-cultural and contextual factors that are associated with travel. For example, Banister et al. (2011) highlight the importance of societal values and aspirations to travel, and the need to 'reconfigure' these, alongside (the more usual focus on) technological solutions, the provision of infrastructure and pricing options, in the search for sustainable travel. The emphasis in this chapter, however, is on the location and form of development and the relationship this might have with sustainable travel, with a view that there are also wider policy areas that need to be progressed. The commentary is also made within the context of increasing urbanization globally: there is much potential for encouraging greater sustainability in travel if the location of homes, workplaces and other activities can be effectively planned, with urban structure providing the 'physical rationale' or the 'envelope' for travel (Hickman et al. 2009). Urban planning can potentially play a major role in contributing to reduced emissions, and also provide increased resilience for urban areas in future times of oil scarcity and higher oil prices. Added attractions are that urban planning as a policy tool is often within the remit of governments and can be viewed as effectively 'free' to the end user (for example, when compared to new vehicle technologies or road pricing, where a cost is incurred to the motorist). Influencing sustainable travel behaviours through an effective design of the built environment can also help facilitate other policy measures, for example, investment in public transport, walking, cycling or smarter travel choice options (such as travel planning or personalized travel planning). Effective urban planning may encourage people out of their cars where the urban structure is currently unsupportive. There are also major co-benefits in terms of the health effects of active travel.

The end objective in integrating urban structure and transport is to move beyond the current practice, which is at best ad hoc, particularly in the non-European context where land use planning systems tend to be weak or even non-existent, and where there are often major difficulties in providing a supportive urban form to public transport, walking and cycling investments. This can be done within the context of improving sustainability in travel, reducing $CO_2$ emissions and also creating attractive areas for living. The central arguments pursued are (1) to rearticulate the role of urban form and layout in setting the 'envelope of possibilities' in travel decision-making and behaviours and (2) to understand the integrated urban planning and transport 'nexus' as part of a wider rationale for sustainable mobility. The transport sector certainly needs to start playing a more significant role in delivering $CO_2$ emission reduction ambitions (Hickman and Banister, 2007a; Ewing et al., 2008; Sperling and Gordon, 2009; Hickman et al., 2010a). Perhaps urban structure also needs to 'step up to the plate'.

## THE EARLY LITERATURE: URBAN STRUCTURE AND TRAVEL

The early debate concerning the potential relationships between urban structure and travel focused on the role of density in supporting the design of public transport systems (Pushkarev and Zupan, 1977), translated more broadly into an international city comparative analysis on density and energy consumption in transport (Newman and Kenworthy, 1989), the thesis being that higher densities were associated with reduced travel distances and transport energy consumption. A proliferation of research followed, covering a wider range of built environment factors, such as density, settlement size, jobs-housing balance, mix of use, the location of development, and extending into local neighbourhood and street design; all of these issues were seen as being associated with travel at various levels of significance. Gradually different metrics of travel (by mode, distance, energy consumption, $CO_2$ emissions), scales of analysis and socio-economic and attitudinal variables were built into the understanding of urban structure and travel relationships (Cervero, 1989, 1996; Breheny and Rookwood, 1993; Frank and Pivo, 1994; Ewing, 1995; Handy, 1995; Naess and Sandberg, 1996; Banister et al., 1997; Kitamura et al., 1997; Headicar and Curtis, 1998; Boarnet and Crane, 1999; Newman and

Kenworthy, 1999; Stead, 2001; Krizek, 2003; Schwanen and Mokhtarian, 2005b; Limtanakool et al., 2006; Hickman and Banister, 2007b; Levinson and Krizek, 2008; Hickman et al., 2009).

The built environment variables were translated into the 3Ds (density, diversity and design) (Cervero and Kockelman, 1997), 5Ds (with the addition of destination accessibility and destination to transit) (Ewing and Cervero, 2001) and even 7Ds (demand management and demographics) (Ewing and Cervero, 2010; Hickman et al., 2010b). Trip frequency was primarily seen as a function of socio-economic characteristics and secondarily the built environment; trip length was seen as primarily a function of the built environment and secondarily of socio-economic characteristics; and mode share and vehicle miles travelled as reliant on both (Ewing and Cervero, 2001). The 'meta analysis' view was that destination accessibility, distance to downtown and neighbourhood design (intersection density and street connectivity) were the most significant built environment variables (Ewing and Cervero, 2010).

There has been, however, much debate, and in parts deep scepticism, as to the scale of the impact of urban form on travel (Gordon and Richardson, 1989, 1995, 1997; Breheny, 1992; Bruegmann, 2006), largely from the neoliberal perspectives of not supporting public intervention in planning, or in deeming that policies to combat urban sprawl are ineffective, or that the public, in the main, wish to live in areas at low density levels, with large houses and gardens, and access to the motor car. Gordon and Richardson have, at least, provided us with some of the best prose in this area, for example:

> The idea of planners turning our lives upside down in pursuit of a single-minded goal is as horrible as it is alien. Newman and Kenworthy's world is the Kafkaesque nightmare that Hayek always dreaded, a world where consumers have no choice, relative prices have no role and planners are tyrants . . . Newman and Kenworthy have written a very troubling paper. Their distortions are not innocent, because the uninformed may use them as ammunition to support expensive plans for central city revitalisation and rail transit projects or stringent land use controls in a futile attempt to enforce urban compactness . . . perhaps Newman and Kenworthy would be well advised to seek out another planet, preferably unpopulated, where they can build their compact cities from scratch with solar powered transit. (Gordon and Richardson, 1989, pp. 344–5)

From a different angle, Breheny and Rookwood (1993) looked beyond the compact city to consider the potential for development at the regional level, conceptualizing the notion of sustainable development as providing higher densities of development around the public transport system while restricting growth in less accessible areas. Hall and Ward (1998) developed this into regional level proposals where development is strategically located in new settlements around stations on existing or new rail lines, for example, in the 'City of Anglia' and 'City of Kent'. Sudjic (1993) considers the development of urban centres as conglomerations spreading over large regions; Castells and Hall (1994) as centres of technological growth or 'technopoles'.

The 'compact city' ideal, and variants of it, such as 'polycentric' growth and 'transit orientated development', have become very influential concepts in policy-making at the governmental level, certainly in Europe and also to a growing extent in North America, Australasia and elsewhere. The Urban Renaissance movement in the UK and New Urbanism in the US enthusiastically took up many of these ideas. Figure 5.1 shows the core principle, as illustrated by Rogers (1997), with higher densities and mix

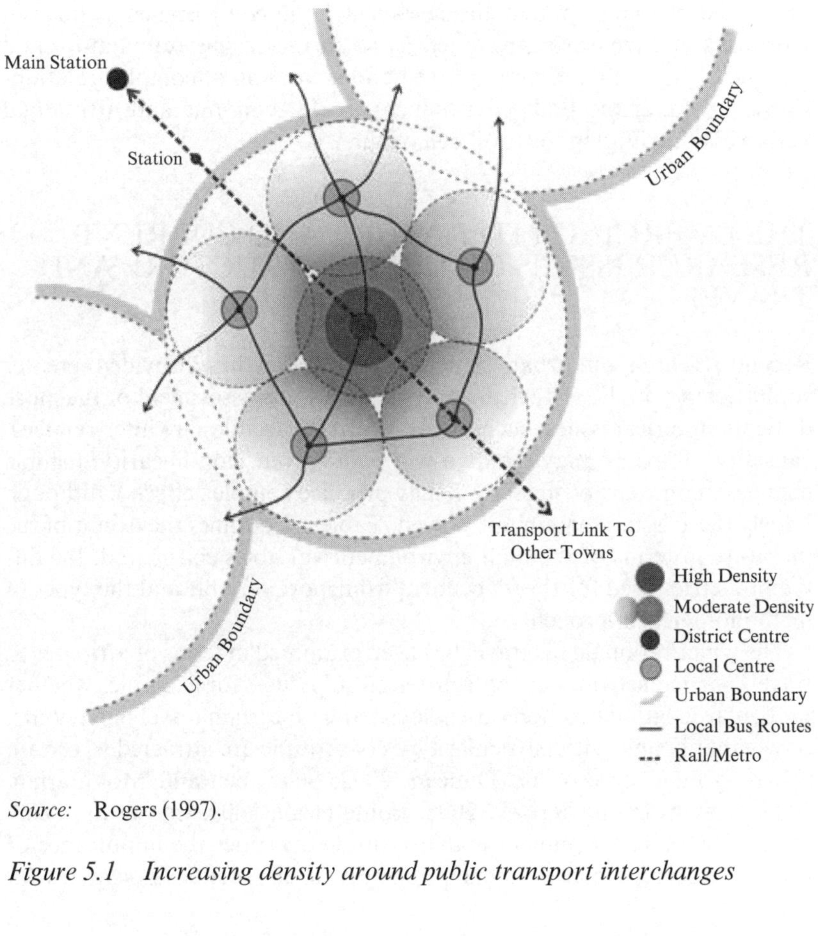


*Source:* Rogers (1997).

*Figure 5.1 Increasing density around public transport interchanges*

of use proposed around public transport interchanges (of various forms: rail, metro, bus rapid transit, bus). Perhaps less impressive has been the actualization of the theory on the ground. Many European cities do show high levels of compaction (Freiburg, Strasbourg, Groningen, Oxford, Cambridge, Paris and London), but even here, beyond the central urban area, most are still fairly dispersed, with low density developments on their edges; and development beyond, in the rural areas, usually dispersed with little strategic coordination. The counterfactual is, however, difficult to know, and certainly when compared to some sprawling North American (Houston and Dallas) or Australasian (Brisbane and Auckland) cities, compaction levels are much greater in Europe, and public transport, walking and cycling levels are much higher. In the developing countries many cities (Lagos, Manila, Bangkok, Mexico City) are in the main unplanned, have sprawled over very large areas and have become very car orientated. Perhaps, though, the caveats highlighted by Breheny, and even Gordon and Richardson, have led us to an increasing examination and critique of the data and an improved understanding of complex relationships, including the built environment, socio-economic and attitudinal variables contributing to travel behaviour.

## THE EMERGING LITERATURE AND CURRENT RESEARCH ISSUES ON URBAN STRUCTURE AND TRAVEL

Recent research on urban structure and travel has provided greater sophistication in the statistical analysis, with some assessment of the more difficult empirical issues, such as co-linearity (variables are inter-related), causality (impacts may be two-way) and even non-linearity (simple changes in one part of the system may produce complex effects), although largely these issues remain unresolved. Table 5.2 outlines the extent of the literature in terms of the built environment variables considered, the different metrics used for the 'dependent' transport variable and the types of methodological approach.

The wider rationale for travel has been examined in terms of attitudes to travel, particularly in view of 'self-selection' issues, for example, whether particular location decisions are associated with certain travel behaviours, or whether people with particular types of attitude are attracted to certain urban forms (Cervero and Duncan, 2002; Schwanen and Mokhtarian, 2005a; Levinson and Krizek, 2008; Bohte et al., 2009; Cao et al., 2009; Naess, 2009). The argument is that attitude overrides the importance of urban structure. Hence, the magnitude of travel benefits associated with

*Table 5.2 Urban form and travel relationships – coverage of the literature*

| Urban structure/ Socio-economic variable | Travel dependent variable | Methodological approach |
|---|---|---|
| • Density<br>• Settlement size<br>• Provision and mix of land use, jobs-housing balance<br>• Location<br>• Regional structure and accessibility<br>• Local street layout, neighbourhood design, parking<br>• Socio-economic characteristics<br>• Attitudinal and cultural characteristics | • Traffic volume<br>• Number of trips<br>• Travel distance<br>• Mode<br>• Car ownership<br>• Journey purpose, e.g. journey to work, non-work travel<br>• Congestion<br>• Energy consumption<br>• CO2 emissions<br>• Electronic social interaction rather than physical travel<br>• Health | • Theoretical argument<br>• Descriptive analysis<br>• Bi-variate analysis, e.g. correlation analysis<br>• Multi-variate analysis, e.g. linear and logistic regression, probit regression, propensity score matching, structural equation modelling<br>• Simulation modelling, including land use and transport integration models<br>• Various datasets, e.g. aggregate and disaggregate level travel surveys<br>• Some limited longitudinal analysis, but mainly cross-sectional data used |

*Source:* Hickman et al. (2009).

compact urban development patterns may be overstated if attitudinal variables are not also considered. A meta analysis of the empirical studies in this area (Cao et al., 2009; Ewing and Cervero, 2010) finds that there is 'resounding evidence' of statistical associations between the built environment and travel behaviour, independent of self-selection influences, that residential self-selection 'attenuates' the effects of the built environment on travel, but that built environment factors are the more significant factors. A further difficulty for researchers is when self-selection is based on variables that are not observed, such as unmeasured attitudes and other social factors (Van Wee, 2009). Boarnet (2006), and others, have also examined the linkages between the built environment and health, largely in terms of travel behaviour and inactive lifestyles. The research finds that other factors, including diet and the availability of processed and fast food, are also contributory explanations for the rise in obesity. However, it seems clear that changes to the built environment are part of the solution. The existence of attitudinal and self-selection issues does not predicate against the building of neighbourhoods that support sustainable travel. The viewpoint from Naess (2009) seems to stand – if people self-select to meet their travel (or other) preferences, it is self-evident that 'urban structure matters'.

There are also wider issues involved. People's choice of home location is part of a complex decision-making process, including family and friendship links, workplace and other activity locations, levels of income (people with higher incomes have a wider set of options to choose from), aspiration and so on (Hickman, 2007). Hence, urban structure is only one part of a complex rationale for travel. For example, it is likely that in areas with high-income house price mismatch (that is, people cannot afford to live where they would like to) that people are 'forced' to live in areas further away from their optimal locations in choice terms. Hence, again, the simplistic built environment and travel relationship becomes cloudy. There are likely to be differences in interaction between variables, including self-selection effects, between very different contexts, for example, in California and the more compact parts of the UK and rest of Europe. The current research does not yet shed light on these contextual issues. There is also little analysis from the developing world or other contexts such as Eastern Europe.

In parallel, other recent research themes note the importance of effective public transport network planning, particularly in suburban and ex-urban areas (Mees, 2010), that is, the extent of integration within and between forms of public transport, drawing on the model of public transport in Zurich. The finding here is that the role of urban planning may not be as important in supporting high levels of public transport in some relatively low density urban areas. There are further angles concerning the influence of electronic interaction on travel and issues of substitution versus complementarity (Mokhtarian, 1990; Cairns, 1995; Castells, 1996; Hall and Pain, 2009) and the effect of social networks on travel (Carrasco and Miller, 2006; Farber and Páez, 2009). Electronic interaction possibilities (physical travel is not always required) add a further layer of complexity to the previous issues of opportunity (there are constraints in access, such as income) and aspiration (the nearest opportunities are not always chosen). Accessing activities may also increasingly not be the prime focus of travel, with travel itself used productively (for example, working or accessing activities while travelling) (Laurier, 2004; Lyons and Urry, 2005; Larsen et al., 2006), or at times carried out for 'affective' emotions such as enjoyment (travel can be more than a derived demand) (Mokhtarian and Salomon, 2001; Anable and Gatersleben, 2005; Lucas et al., 2011).

A number of cultural geographers and sociologists have begun to examine the social and cultural dimensions of travel and mobility, including car use. This emergent 'mobility turn' conceives car use and mobility as more than a movement between an origin and a destination, as part of the sociology and practice of life, with various inter-relating elements (people and means of travel), rules, norms, habits and other factors (Sheller, 2004; Cresswell, 2006; Sheller and Urry, 2006; Urry, 2007; Dennis and Urry, 2009; and see Chapter 2). Hence, the disparate but related fields of research are beginning to explore the complexities of life, and travel within this – and the built environment, in terms of its shape and use, is just one part. Each of these areas are emerging in research terms, with very close links to the urban structure and travel field, and may in the future prove fundamental to our understanding of travel. The urban structure and travel relationship is not static, indeed may be weakening as the 'death of distance' (Cairncross, 1997) potentially becomes more important over time.

## CONCLUSIONS AND IMPLICATIONS FOR FUTURE MOBILITY

The potential for the city to develop as the milieu of innovation is perhaps the principal reason for encouraging urbanization, and urbanization of a certain type – where the result is the vital and vibrant city. There are many types of attractive urbanization, including the conventional concepts of the compact city and polycentricity (see recent study from OECD, 2012). There are even emerging urban forms such as the 'Aerotropolis' (Kasarda and Lindsay, 2011), with new 'centres of activity' developing around some major airport hubs. All should be judged against their outputs in sustainability terms: economic, social and environmental, and also their ability to provide a high quality of life. Within this, urban structure can assist in making travel more sustainable, involving reduced $CO_2$ emissions and other adverse impacts. Transport, as ever, remains low down in the decision-making process in residential and employment locational decisions and city design, and is seen mainly as a facilitator to the effective development and working of the city. Policy-making, in the location and form of housing, employment and other facilities, and investment in transport, however, is critically important – as part of the solution – to achieving a form of urban growth that can enrich life and achieve progress against strategic societal goals. Too much of the common practice in urban and transport planning leads to dispersal and sprawl, a lack of urban character, distinctiveness and vitality, and car dependency and high $CO_2$ emissions from travel. As ever, there is a huge gap between the theory (which is partially disputed) and practice. The emergent developing world can seek to plan its cities in a much more effective manner than at present, with development strategically located to help achieve sustainability aspirations. In many countries, though, this is a very large step. Included here are North and Latin America, and indeed many parts of Europe. This is perhaps the major task within the urbanization phenomenon: how to steer growth – globally across the emerging cities and also retrofitting the historic cities – in a way that helps support sustainable mobility and wider liveability.

The current understanding is that urban form and layout sets the 'envelope of possibilities' for travel (Hickman et al., 2009), and this is modified according to network, socio-economic, attitudinal and other factors. Built environmental factors such as density, location, mix of use and neighbourhood design all prove to be significantly associated with travel. These relationships differ by scale (at the individual and aggregate level) and by context. The location and form of development at the strategic (city, region) and local (neighbourhood design) levels both appear to be important, and it is perhaps the former that is most ignored in terms of potential impacts on travel. The wider rationale for travel is beginning to be understood in terms of the emerging electronic interaction possibilities, and the psychological, cultural and sociological dimensions (see Chapters 2 and 6). Perversely, where the evidence is the most extensive and strongest (North America), the practice is amongst the weakest. Underlying much of the debate has been the confusion in the evidence, with protagonists simply pushing from different angles. As usual, the reality is more complex, and there are multiple forms of urbanization. Much of the debate has been overtaken by the imperative to tackle climate change and the emerging energy supply constraints. The trends in transport in the future will reflect how urban structure is managed, and also the wider issues as discussed. In time, policy levers can be better shaped to influence social and cultural beliefs, alongside the conventional network improvements, travel pricing changes, technological improvements and urban planning dimensions. But, in an era where achieving sustainable lifestyles and lower $CO_2$ emissions in travel have become central policy goals, and where funding constraints often mitigate against major network enhancements in many contexts, the more effective use of spatial planning – in central urban areas and the suburbs – becomes critical.

## REFERENCES

Anable, J. and B. Gatersleben (2005), 'All work and no play? The role of instrumental and affective factors in work and leisure journeys by different travel modes', *Transportation Research Part A: Policy and Practice*, **39**(2–3), 163–81.

Banister, D., S. Watson and C. Wood (1997), 'Sustainable cities, transport, energy and urban form', *Environment and Planning B*, **24**, 125–43.
Banister, D., K. Anderton, D. Bonilla, M. Givoni and T. Schwanen (2011), 'Transport and the environment', *Annual Review of Environment and Resources*, **36**, 247–70.
Boarnet, M. (2006), 'About this issue: planning's role in building healthy cities. An introduction to the special issue', *Journal of the American Planning Association*, **72**, 5–9.
Boarnet, M.G. and R. Crane (1999), *Travel by Design: The Influence of Urban Form on Travel*, New York: Oxford University Press.
Bohte, W., K. Maat and B. van Wee (2009), 'Measuring attitudes in research on residential self selection and travel behaviour: a review of theories and empirical research', *Transport Reviews*, **29**, 325–57.
Breheny, M. (1992), 'The contradictions of the compact city: a review', in M. Breheny (ed.), *Sustainable Development and Urban Form*, London: Pion, pp. 138–59.
Breheny, M. and R. Rookwood (1993), 'Planning the sustainable city region', in A. Blowers (ed.), *Planning for a Sustainable Environment*, London: Earthscan.
Bruegmann, R. (2006), *Sprawl: A Compact History*, Chicago, IL: University of Chicago Press.
Cairncross, F. (1997), *The Death of Distance: How the Communications Revolution Will Change Our Lives*, Boston, MA: Harvard Business School Press.
Cairns, S. (1995), *Travel for Food Shopping: The New Doctrine and Some Alternatives*, Transport Studies Unit, Oxford: University of Oxford.
Cao, X., P.L. Mokhtarian and S.L. Handy (2009), 'Examining the impacts of residential self-selection on travel behaviour: a focus on empirical findings', *Transport Reviews*, **29**, 359–95.
Carrasco, J. and E. Miller (2006), 'Exploring the propensity to perform social activities: a social network approach', *Transportation*, **33**, 463–80.
Castells, M. (1996), *The Rise of the Network Society*, Oxford: Blackwell.
Castells, M. and P. Hall (1994), *Technopoles of the World: The Making of 21st Century Industrial Complexes*, London: Routledge.
Cervero, R. (1989), 'Jobs-housing balancing and regional mobility', *Journal of the American Planning Association*, **55**, 136–50.
Cervero, R. (1996), 'Jobs-housing balancing revisited', *Journal of the American Planning Association*, **62**, 492–511.
Cervero, R. and M. Duncan (2002), 'Residential self selection and rail commuting: a nested logit analysis', working paper, University of California, Berkeley, California.
Cervero, R. and K. Kockelman (1997), 'Traffic demand and the 3Ds: density, diversity, and design', *Transportation Research Part D*, **2**(3), 199–219.
Climate Change Committee (2010), *The Fourth Carbon Budget. Reducing Emissions through the 2020s*, London: Climate Change Committee.
Cresswell, T. (2006), *On the Move: Mobility in the Modern Western World*, London: Routledge.
Dennis, K. and J. Urry (2009), *After the Car*, Cambridge: Polity Press.
Ewing, R. (1995), 'Beyond density, mode choice, and single-purpose trips', *Transportation Quarterly*, **49**, 15–24.
Ewing, R. and R. Cervero (2001), 'Travel and the built environment. A synthesis', *Transportation Research Record*, **1780**, 87–114.

Ewing, R. and R. Cervero (2010), 'Travel and the built environment. A meta analysis', *Journal of the American Planning Association*, **76**, 265–94.
Ewing, R., K. Bartholomew, S. Winkelman, J. Walters and D. Chen (2008), *Growing Cooler: The Evidence on Urban Development and Climate Change*, Washington, DC: Urban Land Institute.
Farber, S. and A. Páez (2009), 'My car, my friends, and me: a preliminary analysis of automobility and social activities', *Journal of Transport Geography*, **17**, 216–25.
Frank, L. and P. Pivo (1994), 'Impacts of mixed use and density on utilization of three modes of travel: single-occupant vehicle, transit, and walking', *Transportation Research Record*, **1466**, 44–52.
Glaeser, E. (2011), *Triumph of the City*, London: Macmillan.
Gordon, P. and H. Richardson (1989), 'Gasoline consumption and cities: a reply', *Journal of the American Planning Association*, **55**, 342–5.
Gordon, P. and H. Richardson (1995), 'Sustainable congestion', in J. Brotchie, M. Batty, E. Blakely, P. Hall and P. Newton (eds), *Cities in Competition: Productive and Sustainable Cities for the 21st Century*, Melbourne: Longman, pp. 348–58.
Gordon, P. and H. Richardson (1997), 'Are compact cities a desirable planning goal?', *Journal of the American Planning Association*, **63**, 95–106.
Hall, P. (1988), *Cities of Tomorrow: An Intellectual History of Urban Planning and Design in the Twentieth Century*, Oxford: Basil Blackwell.
Hall, P. (1998), *Cities in Civilization: Culture, Innovation, and Urban Order*, London: Weidenfeld and Nicolson.
Hall, P. and K. Pain (2009), *The Polycentric Metropolis: Learning from Mega-city Regions in Europe*, London: Earthscan.
Hall, P. and C. Ward (1998), *Sociable Cities: The Legacy of Ebenezer Howard*, Chichester: John Wiley.
Handy, S. (1995), *Understanding the Link Between Urban Form and Travel Behaviour*, Washington, DC: Transportation Research Board.
Headicar, P. and C. Curtis (1998), 'The location of new residential developments: its influence on car-based travel', in D. Banister (ed.), *Transport Policy and the Environment*, London: Spon.
Hickman, R. (2007), 'Reducing the need to travel. A micro analysis of new housing location and the commute to work', Unpublished PhD thesis, University College London.
Hickman, R. and D. Banister (2007a), 'Looking over the horizon: transport and reduced $CO_2$ emissions in the UK by 2030', *Transport Policy*, **14**, 377–87.
Hickman, R. and D. Banister (2007b), 'Transport and reduced energy consumption: what role can urban planning play?', working paper no. 1026, University of Oxford TSU.
Hickman, R., C. Seaborn, P. Headicar and D. Banister (2009), *Planning for Sustainable Travel: Summary Guide*, London: Halcrow Group and Commission for Integrated Transport.
Hickman, R., O. Ashiru and D. Banister (2010a), 'Transport and climate change: simulating the options for carbon reduction in London', *Transport Policy*, **17**, 110–25.
Hickman, R., C. Seaborn, P. Headicar and D. Banister (2010b), 'Planning for sustainable travel: integrating spatial planning and transport', in M. Givoni and D. Banister (eds), *Integrated Transport: From Policy to Practice*, London: Routledge, pp. 33–54.

Kasarda, J. and G. Lindsay (2011), *Aerotropolis: The Way We'll Live Next*, New York: Farrar, Straus and Giroux.
Kitamura, R., P. Mokhtarian and L. Laidet (1997), 'A micro-analysis of land use and travel in five neighbourhoods in the San Francisco Bay Area', *Transportation*, **24**, 125–58.
Krizek, K. (2003), 'Residential re-location and changes in urban travel. Does neighbourhood scale urban form matter?', *Journal of the American Planning Association*, **69**(3), 265–81.
Larsen, J., K. Axhausen and J. Urry (2006), 'Geographies of social networks: meetings, travel and communications', *Mobilities*, **1**, 261–83.
Laurier, E. (2004), 'Doing office work on the motorway', *Theory, Culture and Society*, **21**, 261–77.
Levinson, D. and K. Krizek (2008), *Planning for Place and Plexus. Metropolitan Land Use and Transport*, New York: Routledge.
Limtanakool, N., M. Dijst and T. Schwanen (2006), 'The influence of socio-economic characteristics, land use and travel time consideration on mode choice for long-distance trips', *Journal of Transport Geography*, **14**, 327–41.
Lucas, K., E. Blumenberg and R. Weinberger (2011), *Auto Motives: Understanding Car Use Behaviours*, Bingley: Emerald.
Lyons, G. and J. Urry (2005), 'Travel time use in the information age', *Transportation Research Part A: Policy and Practice*, **39**, 257–76.
Mees, P. (2010), *Transport for Suburbia: Beyond the Automobile Age*, London: Earthscan.
Mokhtarian, P. (1990), 'A typology of relationships between telecommunications and transportation', *Transportation Research Part A: General*, **24**, 231–42.
Mokhtarian, P. and I. Salomon (2001), 'How derived is the demand for travel? Some conceptual and measurement considerations', *Transportation Research Part A: Policy and Practice*, **35**, 695–719.
Naess, P. (2009), 'Residential self selection and appropriate control variables in land use: travel studies', *Transport Reviews*, **29**, 293–324.
Naess, P. and S. Sandberg (1996), 'Workplace location, modal split and energy use for commuting trips', *Urban Studies*, **33**, 557–80.
Newman, P. and J. Kenworthy (1989), *Cities and Automobile Dependence: An International Sourcebook*, Aldershot: Gower.
Newman, P. and J. Kenworthy (1999), *Sustainability and Cities: Overcoming Automobile Dependence*, Washington, DC: Island Press.
OECD (2012), *Compact City Policies: A Comparative Assessment*, June, Paris: OECD.
Pushkarev, B. and J. Zupan (1977), *Public Transportation and Land Use Policy*, Bloomington, IN: Indiana University Press.
Rogers, R. (1997), *Cities for a Small Planet*, London: Faber.
Schwanen, T. and P. Mokhtarian (2005a), 'What affects commute mode choice: neighborhood physical structure or preferences toward neighborhoods?', *Journal of Transport Geography*, **13**, 83–99.
Schwanen, T. and P. Mokhtarian (2005b), 'What if you live in the wrong neighborhood? The impact of residential neighborhood type dissonance on distance traveled', *Transportation Research Part D: Transport and Environment*, **10**, 127–51.
Sheller, M. (2004), 'Automotive emotions: feeling the car', *Theory, Culture and Society*, **21**, 221–42.

Sheller, M. and J. Urry (2006), 'The new mobilities paradigm', *Environment and Planning A*, **38**, 207–26.
Sperling, D. and D. Gordon (2009), *Two Billion Cars. Driving Toward Sustainability*, New York: Oxford University Press.
Stead, D. (2001), 'Relationships between land use, socioeconomic factors and travel patterns in Britain', *Environment and Planning B: Planning and Design*, **28**, 499–528.
Sudjic, D. (1993), *The 100 Mile City*, London: Flamingo.
Urry, J. (2007), *Mobilities*, Cambridge: Polity Press.
Van Wee, B. (2009), 'Location choices, travel behaviour and transport externalities', *Transport Reviews*, **29**, 279–92.
World Bank (2010), *World Development Indicators. Country Data*, Washington, DC: World Bank.

# 6. Unpacking travel behaviour

**Malek Al-Chalabi**

## INTRODUCTION

The purpose of this chapter is to present the reader with an overview of what behaviour means in the context of travel by providing an overview of the relevant academic developments that aim to unpack travel behaviour; highlight notable theoretical and conceptual frameworks; illustrate key points with relevant case studies; and discuss the implications of these findings for behavioural change research. In order to do so, the chapter is structured as follows. The first section provides an overview of the relevant developments that have aimed to unpack travel behaviour, giving consideration to the traditional approaches and how they have evolved. The second section highlights the relevant theoretical and conceptual frameworks of travel behaviour, noting the advantages, disadvantages and drivers to behaviour change from each framework and how they compare to one another. The third section provides relevant case studies that illustrate the points made in the previous sections. The final section discusses the implications of these findings for behavioural change research, drawing from recent works in behavioural economics and habit research.

## UNPACKING TRAVEL BEHAVIOUR

Energy use in transport has increased substantially over the recent past (Banister and Banister, 1983; Price and Probert, 1995; Hickman and Banister, 2007 and also Chapter 8), and travel is an integral part of individual energy consumption. Therefore, understanding travel demand has been of interest to a myriad of researchers. The means by which researchers have attempted to unpack travel behaviour is presented by examining the trip- and activity-based approaches as well as highlighting how travel behaviour can be conceptualized in the short, medium and long term. These distinctions are important because they illustrate the many facets of travel behaviour.

## Trip- and Activity-based Approaches

The trip-based approach and the activity-based approach have both been used to unpack travel behaviour, and the former approach has traditionally dominated the academic literature.

Individual trips are the unit of analysis for the trip-based approach and this is exemplified in the four-step model that aims to forecast future demand and performance of transport systems. The four-step model includes the following four steps: trip generation, trip distribution, mode choice and assignment (McNally and Rindt, 2008). The first step, trip generation, estimates the number of person trips (home-based and non-home-based trips) produced from (and attracted to) each zone in the area of interest. The second step, trip distribution, establishes the number of trip interchanges, which is defined as the number of trips from one zone to the other. The third step, mode choice, splits the person trips by travel mode, which then presents the number of vehicle trips and number of transit trips. The final step, assignment, allocates the vehicle trips to the roadway network to present link volumes, travel times and person trips (Bhat and Koppelman, 1999). This approach has been criticized for several reasons, because there are no linkages between trips and activities, it does not incorporate the element of time constraints, nor does it take into account underlying activities or interpersonal linkages (McNally, 2000). Critiques of the trip-based approach may also be found in Kitamura (1988). This has led to an interest in the activity-based approach – which takes a broader and more comprehensive approach towards understanding travel.

The activity-based approach is distinguished from the trip-based approach because it views travel as a series of trips made with linkages and sequences. The activity-based approach also conceptualizes time as an entity that is continuous and an entity that is embedded in individual activity/travel decisions (Kurani and Lee-Gosselin, 1996). Therefore, some would argue that the framework is not only able to evaluate the linkages between trips, but can also provide a more holistic approach to understanding travel.

## Conceptualizing Travel Behaviour: Short-, Middle- and Long-term Decisions

Building on the work of Salomon and Ben-Akiva (1983), Schwanen and Mokhtarian (2007) developed a conceptual scheme that relates various travel decisions and terms (short, medium and long) (Figure 6.1).

What this conceptual model demonstrates is that travel behaviour has

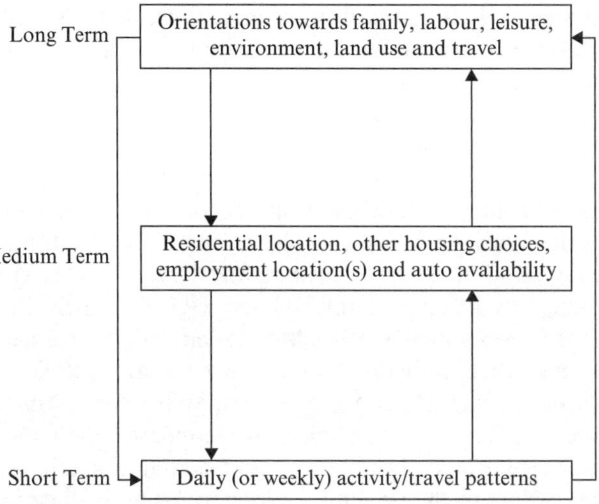


*Source:* Schwanen and Mokhtarian (2007).

*Figure 6.1 Travel behaviour and short-, middle- and long-term decisions*

various aspects and influences. Day to day (or week to week) activity/travel pattern decisions are partly rooted in other decisions, like residential location, employment location and lifestyle orientation (some of which are evolving decisions over time). These aspects are important to consider because they highlight two important points: many different aspects of travel behaviour can be changed; and there are, therefore, different ways to change travel behaviour. Noting that there are different aspects associated with travel behaviour, it is no coincidence that there are different theoretical and conceptual frameworks premised on different underlying assumptions that aim to describe travel behaviour. Understanding and examining this literature is the focus of the next section.

## THEORETICAL AND CONCEPTUAL FRAMEWORKS

As explained earlier, there are different aspects associated with travel behaviour, and subsequently, there are different theoretical and conceptual frameworks that are premised on different underlying assumptions. Notable contributions to the travel behaviour literature include utility theory and relevant extensions of utility theory, the socio-psychological literature and the time geography literature. After a brief overview of the different approaches, a comparison is made towards their contribution to understanding travel behaviour and understanding important drivers for behaviour change.

## Utility Theory and Extensions

Traditionally, utility models based on neoclassical economics have been used from the 1960s onwards to evaluate, understand and model human choice behaviours in travel (Schwanen and Lucas, 2011). Most of them are enlightened by Random Utility Theory (RUT), which was developed by McFadden (see Domencich and McFadden, 1975). The theory is based on the premise that individuals rank the alternative choices they have based on utility, which is defined as the capability for alternative choices to meet their desires of each choice. Subsequently, the choice with the highest utility is selected. The theory assumes that individuals make comparisons according to the principle of least effort and allocation of scarce resources (Schwanen and Lucas, 2011). A comprehensive review of the developments and improvements of RUT since its initial development can be found elsewhere (McFadden, 2001), but the major developments are highlighted here.

Since its original development, RUT approaches have begun to incorporate other factors (Swait, 2001) and they have been extended in a variety of ways (see Schwanen and Lucas, 2011 for an overview). The reasons for incorporating additional factors were because of the limitations inherent in RUT, which include: decisions about travel are interdependent with other choices; people's travel preferences can vary in places and times; human beings are not perfectly informed about their choices; RUT focuses on the instrumental aspects more than the emotional, symbolic and cultural aspects; decision making is adaptive over time, and people form habits and, therefore, their decisions may then not go through a conscious process of weighing the costs and benefits. Some important extensions of RUT have included: challenging the assumption that humans are perfectly informed about their choices and examining uncertainty/people's risk attitudes with the decision to travel; accounting for symbolic, affective and social motivations for using a car by including variables that measure emotions, attitudes and normative beliefs (Koppelman and Lyon, 1981); Domarchi et al., 2008); and attempting to capture habits (Weinberger and Goetzke, 2011).

The limitations inherent in RUT demonstrate, amongst other things, the importance of psychological factors in travel decision making. Contributions from the socio-psychological literature towards understanding travel behaviour are presented in the next section.

**Socio-psychological Theories**

A comprehensive literature review of the socio-psychological developments of travel behaviour can be found in Bamberg et al. (2011). Here, three prevalent socio-psychological theories revolving around travel are introduced: the theory of planned behaviour (Ajzen, 1991), the theory of interpersonal behaviour (Triandis, 1977) and the value belief norm theory (Stern et al., 1999), which stems from the norm activation theory (Schwartz, 1977). After briefly introducing these theories, differences within the literature are highlighted and some recent extensions illustrated.

The theory of planned behaviour investigates how behaviours are formed and linked with intentions. The theory states that there are three elements that determine intentions and behaviour (Figure 6.2). The first element is the attitude towards the behaviour. Attitudes are defined as evaluations (positive or negative) of the behaviour of interest. The second is subjective norms, which are defined as social pressures to perform or not perform certain behaviours. The third element is perceived behavioural

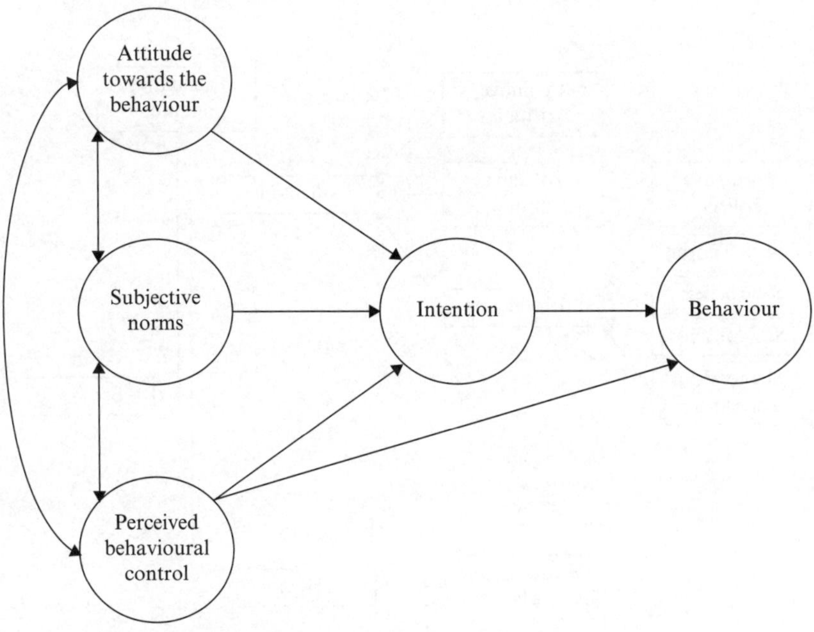


*Source:* Ajzen (1991)

*Figure 6.2 The theory of planned behaviour*

control, which is defined as the difficulty and feasibility of using alternatives based on previous experiences as well as anticipated obstacles. Given these three elements, the theory is based on the premise that intentions to perform behaviours can be therefore predicted with a high accuracy – meaning that intention is an antecedent of behaviour (Ajzen, 1991).

A key question that arises is what happens when a certain type of behaviour (like car driving) is performed frequently over time, and whether attitudes, intentions and behaviour all seem to take place simultaneously. The second theory of interest, the theory of interpersonal behaviour, states that 'as behaviour becomes more frequently performed, past behaviour (or habit) becomes increasingly predictive of behaviour while intentions become less predictive' (Jackson et al., 2003, p. 122). This seems in part to address this question, but Schwanen and Lucas (2011) have identified key differences between these two theories, which seem to be at odds with each other.

The theory of interpersonal behaviour proposes that habit influences people's behaviour, especially when they occur regularly. Therefore, recognizing and appreciating habit, which has the ability to lock in certain behaviours, is critical in order to understand behaviour and to change behaviour (Figure 6.3).

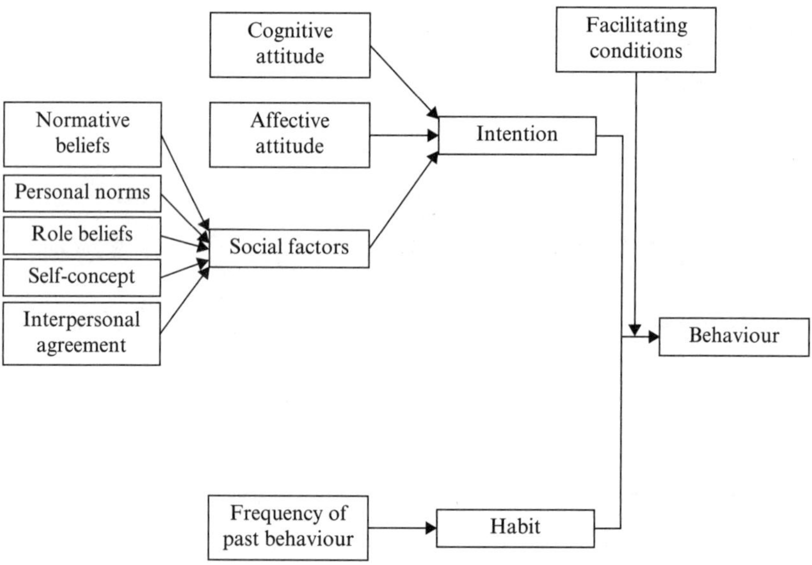


*Source:* Schwanen and Lucas (2011), based on Triandis (1977).

*Figure 6.3 Understanding behaviour based on intention and habits*

As Figure 6.3 illustrates, the theory of interpersonal behaviour encompasses a broader understanding of personal behaviour and incorporates habit as well as intention in order to explain behaviour. Furthermore, social factors and physical constraints (termed facilitating conditions) have a broader conceptualization in the theory of interpersonal behaviour than in the theory of planned behaviour. This provides a richer context towards understanding what options are feasible or not to change behaviour (Stern, 2000).

Alternatively, value-belief norm theory is a theory that uses a causal chain of variables that lead to behaviour (Figure 6.4). It proposes personal norms, defined as the intrinsic obligation to make one's behaviour in line with one's internalized self-standards (Biel and Thøgersen, 2007), as the antecedent of behaviour (instead of intentions – as illustrated in the theory of planned behaviour). Since these can be influenced by society, value-belief norm theory has the promising attribute of taking these factors into account and is, in theory, more understanding of the social nature of behaviours (Bamberg et al., 2011). It is important to note that social norms are defined as appropriate behavioural standards in specific social reference groups, rather than social pressures.

While the three theories presented are socio-psychological, their interpretation of the factors that influence behaviour can at times be quite different. The socio-psychological literature on travel behaviour has a heterogeneous interpretation of what variables influence behaviour. Recently, a series of more conceptual developments have been initiated to integrate

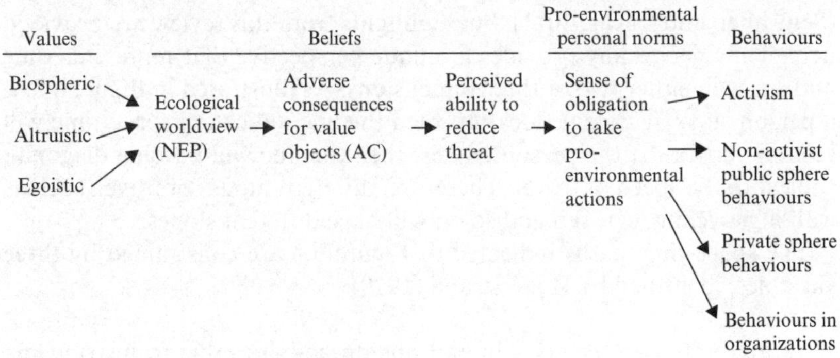


*Note:* NEP – New Environmental Paradigm.

*Source:* Stern (2000).

*Figure 6.4 The value-belief norm theory*

the three theoretical frameworks outlined above so that the limitations inherent in each of them can be addressed and the socio-psychological approach made more consistent (Gardner and Abraham, 2010; Eriksson and Forward, 2011; Ogilvie et al., 2011). One of these developments, the Comprehensive Action Determination Model, highlights the benefits of an integrative framework approach.

The Comprehensive Action Determination Model is based on the theory of planned behaviour, yet is distinguished by incorporating intentional, situational, normative and habitual influences, which the theory of planned behaviour does not (Klöckner and Blöbaum, 2010). The following findings emerged when exploring travel mode choice: most variation in travel mode choice was due to subjective and objective situational constraints (accessibility to alternative modes of transport, cost and so on), yet habits and intentions did have a significant impact on travel mode choice; the model results compared quite well to the theory of planned behaviour and to other less conventional yet mainstream models, but it was able to describe reasons for travel mode variation in finer detail; and habit and intention were mediated by personal norms and social influence, while intention and behaviour were moderated by habits. Other research, based on the theory of planned and other socio-psychological approaches, has supported these findings (Chen and Chao, 2011).

## Time Geography Approaches

The final approach has originated in the accessibility and time geography literature. A more comprehensive overview can be found elsewhere (Schwanen and Lucas, 2011), but highlights from this review are provided here. Time geography provides a unique perspective that represents time and space in either two or three dimensions. As illustrated in Figure 6.5, if a person stays at a single location then the line indicating space-time will remain vertical. If the person moves, then the line will become diagonal, subject to the speed of travel. Therefore, different modes of travel, such as walking, cycling, car, rail and so on will have different slopes.

The space-time paths indicated in Figure 6.5 are constrained by three variables, identified by Hagerstrand (1970):

1. Capability constraints – limitations on activities due to instruments available (maximum speed, for example) as well as biological needs (sleeping, eating and so on).
2. Coupling constraints – times when individuals have an imperative to join others to meet, produce, consume and/or transact – an ever evolving constraint with the development of technology.

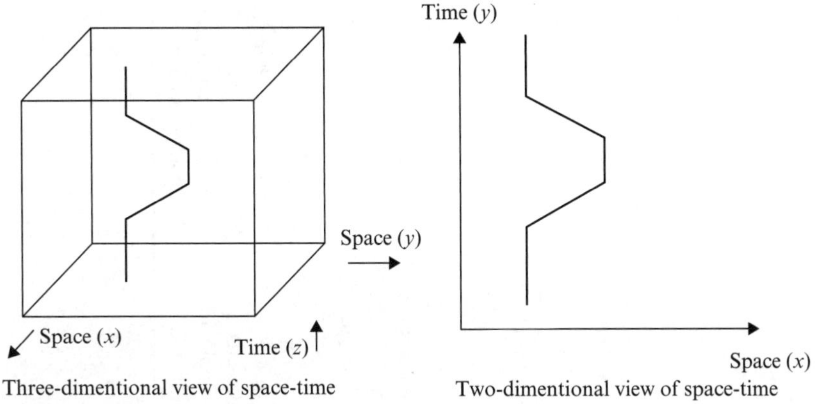


*Source:* Schwanen and Lucas (2011).

*Figure 6.5 Illustration of movement over space-time*

3. Authority/steering constraints – norms, regulations, laws and so on that indicate designated areas are accessible at specific times, such as opening hours.

This perspective can demonstrate whether people have choices in terms of travel alternatives and evaluate how space-time constraints influence travel decisions. The time geography perspective can show that some travellers do not have as many alternatives as advertised because of inherent time-space constraints. This can illustrate, for example, when the car is the only realistic mode of transport one can use within time-space constraints.

**Summary**

As highlighted above, different perspectives (utility theory, socio-psychological theories and time geography) provide different theoretical and conceptual frameworks that are premised on different underlying assumptions – indicating different drivers and variables for behaviour change. A summary of the points highlighted above is given in Table 6.1.

The main approaches that aim to unpack travel behaviour represent certain and different categorized perspectives. Therefore, drivers to change and ways to change behaviour, as highlighted above, are inherently different depending on the approach. This can often lead to an underestimation of behaviour change and the complexities involved in

*Table 6.1 Summary of the main approaches presented in this review*

| Approach | Key theories | Advantages | Weaknesses | Drivers for change[a] |
|---|---|---|---|---|
| Utility theory | Random Utility Theory (RUT) | Provides a strong perspective for viewing alternative choices and evaluating their costs and benefits<br>Recent developments have started to incorporate perception, norms etc. | Human beings are assumed to be perfectly informed about their choices<br>Utility is not the only factor in decision making | Prices<br>Land use<br>Infrastructure improvement |
| Socio-psychological theories | Theory of planned behaviour<br>Theory of interpersonal behaviour<br>Value-belief norm theory | Examines role and formation of travel habits<br>Norms, attitudes and intentions are theoretically evaluated<br>External conditions (social influence) are garnering increased attention | Models are usually not dynamic<br>Findings can be limited to specific samples of the population and can be difficult to generalize | Information<br>Social norms<br>Personal norms |
| Time geography | Time geography | Illustrates space-time constraints and realistic alternatives<br>Highlights what transport modes can/cannot be used as substitutes | Strong emphasis on the physical terrain (infrastructure, design etc.) and less on 'soft' transport policy | Land use<br>Infrastructure changes<br>Time/space changes<br>Institutional rhythms (opening hours) |

*Note:* [a] Tools that can be used to change behaviour.

*Source:* Based on Schwanen and Lucas (2011).

changing behaviour. These theoretical points can be highlighted even further with relevant case studies, and this is the purpose of the next section of the chapter.

## RELEVANT CASE STUDIES

Travel behaviour can be conceptualized as being short, medium or long term, and travel behaviour can be approached from a utility theory perspective, a socio-psychological perspective or a time geography perspective. In order to demonstrate the differences, case studies are presented, each illustrating one of the three approaches discussed and each representing a different time horizon. The first involves providing a free month-long travel card (short term – socio-psychological perspective), the second covers the role of previous experiences in determining car ownership and residential location (medium term – utility perspective) and the third evaluates spatial configurations and quality of life in travel (medium/long term – time geography perspective).

### A Free Month-long Travel Card

Price promotion is consistently used in commercial marketing because it has the potential to disrupt the consumer's regular habits and to consider a new offer, which can change behaviour (Thøgersen, 2009). In a Danish study Thøgersen (2009) aimed to evaluate whether free month-long travel cards could encourage people to leave their cars and use alternative modes of public transport. Participants were randomly selected from a sample population in Copenhagen and mailed a free month-long travel card valued at 995 DKK or 133 euros at the time of publication. The study found that the price promotion led to an increased use of public transport by the participants. Before the intervention, 5 per cent of the commuting trips were by public transport. After the trial the level increased to 10 per cent in the short term and 7 per cent in the longer term (six months later). Similar findings were found in other studies, for example, by Fujii and Kitamura (2003).

The study concluded that the free month-long travel card offer was effective in persuading drivers to try alternative forms of public transport and to change ingrained habits. The experiment may have led to a positive change in the drivers' beliefs about public transport, which made them use it even after the promotion expired. An important indirect effect was that the price promotion strengthened the intentions of the drivers to use public transport, and that the impact of the promotion depended on perceived behavioural control. The author notes that one of the limitations is the assumption that structural conditions, such as frequency of service, route plan, the built environment and so on, are not prohibitive. The accessibility and time geography literature clearly indicate that this is not a valid assumption. It highlights some of the differences between the socio-psychological and time geography approaches. This is important to note because under the right conditions and policy packages, comprehensive interventions can be designed to take these factors into account so that they work constructively together rather than against each other. These points are discussed in greater detail in the chapter's concluding paragraphs.

### Role of Previous Experiences in Determining Car Ownership and Residential Location

The utility approach has many strengths but also some shortcomings, which some scholars try to address by integrating utility theory with other frameworks. Weinberger and Goetzke's (2011) study adopted a random utility modelling approach towards understanding car ownership, but distinguished from the conventional approach by incorporating habit and social norms as well as other social aspects. They did this by analysing two cases. The first case evaluated choices of car ownership and residential location by using a joint probability model and the second case examined peer pressure and social network effects by using a two-stage model that incorporated social influences. The study was carried out in neighbourhoods in New York, US.

The findings were that: attitudes and preferences towards travel behaviour and travel decisions are influenced by past options and experiences; travel decisions are also influenced by one's peers, and this in turn leads to behaviours and subsequently habits formed by social interactions. This means that travel decisions encompass more than just the cost and speed of options, the built environment, infrastructure and so on. The authors also note the importance of social spillovers, such as raising awareness, and promoting certain policy objectives. For example, if more cycling facilities are available, more people will ride and more people will also be exposed to the idea of riding a bicycle.

### Evaluating Spatial Configurations and Quality of Life in Travel

The final case study investigates how spatial configurations and transport facilities could be used to promote efficient mobility from an environmental and spatio-temporal perspective. Ritsema van Eck et al. (2005) define efficient mobility as one that reduces car use and encourages alternative modes of transport. In their study they used three quality of life indicators: feasibility of activity programmes (measured by the extent to which people can maintain their current activity patterns in different spatial configurations); travel time; and travel distance efficiency.

The simulation models used in the study indicate that concentrating activities around areas of public transport could be an effective measure to reduce the use of the car as well as increase the use of public transport. However, this comes with a set of precautions. Pedestrians and cyclists might encounter problems because average distances from home to the activities could increase, and this can be a problem especially for senior citizens who rely more on slow transport modes. Also, those with tight time budgets could face difficulties in realizing their activity programmes if they cannot reach activities by public transport. The study ran simulations for scenarios such as less concentrated and more dispersed areas. The finding, as expected, is that it is generally easier to visit activities in a dense area where distances are shorter than in a dispersed area where distances are greater. Ritsema van Eck et al. (2005, p. 133) conclude that 'spatial policy can gain from a more lifestyle oriented approach'.

A limitation of the approach taken, noted by the authors, is that it assumes activities and activity times are fixed. Thus, the study might have underestimated actual choice and real opportunities since it does not account for flexible working hours, the allocating of tasks amongst household members and the choice of activities within a given time horizon. While there is other research that does take these into account, this highlights yet another difference between the approaches.

## IMPLICATIONS FOR BEHAVIOURAL CHANGE RESEARCH

This chapter has highlighted the following points. First, the main approaches that aim to unpack travel behaviour can be viewed from different theoretical perspectives. Therefore, and second, drivers to change travel behaviour are subsequently perceived differently depending on the perspective and adopted approach. Third, integrative frameworks can be structured to overcome some of the shortcomings of the different approaches in order to better understand and create opportunities to influence travel behaviour. Two examples of research on habit and research in behavioural economics demonstrate the potential of using integrative frameworks.

With respect to the research on habit, it is important to note that this body of research is diversified and conceptualized in a variety of ways.[1] Building on the habit literature, Schwanen et al. (2012) suggest that instead of conceptualizing habit as being automatically cued, individualistic and linked to one's self, it should instead be conceptualized as a tendency brought through repetition, non-individualistic (an emergent property of body-mind assemblages) and linked to the society and community. Furthermore, they suggest that 'widespread, durable behaviour change is unlikely to result from the displacement of automaticity by reasoned action alone but instead demands changes in collective customs' (Schwanen et al., 2012, p. 522). This is relevant to behaviour change as this is the type of integrative thinking that can be used to better understand and appreciate travel behaviour and change.

Turning to behavioural economics, Metcalfe and Dolan (2012) examine the use of behavioural economics in transport and highlight one major assumption, which is that the human mind is conceptualized in two parts: the reflective and the automatic. The authors note that exploring this further is a relevant area of research as it can help improve understanding travel decisions and mobility choices from an integrative perspective. Other scholars have arrived at similar conclusions. Avineri (2012) highlights that current economic thinking assumes that individuals are rational, consistent and make efficient choices, while behavioural economists note that choices in reality may differ from what a rational individual may do because of different contexts. Avineri (2012) also notes that behavioural economics has a role to play in designing behaviour change measures that current economic thinking may overlook. Future areas of research include: critically evaluating human behaviour in decision-making travel scenarios within economic and financial contexts; investigating the contextual effects of individual perceptions and choices on travel alternatives; and understanding and explaining heterogeneity in travel choices.[2]

Research in behavioural economics and habit demonstrate the possibility of examining travel behaviour from a comprehensive perspective, similar to the integrative approaches highlighted earlier. These research strands can then be designed to improve and structure integrative policy packages, where interventions acknowledge the strengths and weaknesses of the different approaches and feed towards a collective goal, rather than remain categorized where they could potentially counteract each other. These types of integrative approaches are needed towards a comprehensive low carbon transition, and one that acknowledges the importance of the social, cultural, physical, economic and geographical spheres collectively.

## ACKNOWLEDGEMENT

The author would like to thank Tim Schwanen for his thoughts and feedback on this chapter.

## NOTES

1. Please refer to the work of Schwanen et al. (2012) for a review of habit research and to Verplanken (2008) for research that integrates habit with behaviour change.
2. Some research has examined the varying role of attitudes in travel decisions. See the work of Prillwitz and Barr (2011).

## REFERENCES

Ajzen, I. (1991), 'The theory of planned behavior', *Organizational Behavior and Human Decision Processes*, **50**, 179–211.

Avineri, E. (2012), 'On the use and potential of behavioural economics from the perspective of transport and climate change', *Journal of Transport Geography*, **24**, September, 512–21.

Bamberg, S., S. Fuji, M. Friman and T. Garling (2011), 'Behaviour theory and soft transport policy measures', *Transport Policy*, **18**, 228–35.

Banister, C. and D. Banister (1983), 'Transport, travel and energy in the UK – trend analysis of published statistics', *Energy Policy*, **11**(1), 39–51.

Bhat, C. and F. Koppelman (1999), 'Activity-based modeling of travel demand', in R. Hall (ed.), *Handbook of Transportation Science*, Norwell: Kluwer Academic Publishers, pp. 35–61.

Biel, A. and J. Thøgersen (2007), 'Activation of social norms in social dilemmas: a review of the evidence and reflections on the implications for environmental behaviour', *Journal of Economic Psychology*, **28**, 93–112.

Chen, C. and W. Chao (2011), 'Habitual or reasoned? Using the theory of planned behavior, technology acceptance model, and habit to examine switching intentions toward public transit', *Transportation Research Part F*, **14**, 128–37.

Dijst, M. and V. Vidakovic (2000), 'Travel time ratio: the key factor of spatial reach', *Transportation*, **27**, 179–99.

Domarchi, C., A. Tudela and A. Gonzalez (2008), 'Effects of attitudes, habit, and affective appraisal on mode choice: an application to university workers', *Transportation*, **35**(5), 585–99.

Domencich, T.A. and D. McFadden (1975), *Urban Travel Demand: A Behavioural Analysis*, Amsterdam: North-Holland.

Eriksson, L. and S.E. Forward (2011), 'Is the intention to travel in a pro-environmental manner and the intention to use the car determined by different factors?', *Transportation Research Part D*, **16**, 372–6.

Fujii, S. and R. Kitamura (2003), 'What does a one-month free bus ticket do to habitual drivers? An experimental analysis of habit and attitude change', *Transportation*, **30**(1), 81–95.

Gardner, B. and C. Abraham (2010), 'Going green? Modeling the impact of environmental concerns and perceptions of transportation alternatives on decisions to drive', *Journal of Applied Social Psychology*, **40**(4), 831–49.
Hagerstrand, T. (1970), 'What about people in regional science?', *Regional Science Association Papers*, **24**(1), 7–21.
Hickman, R. and D. Banister (2007), 'Looking over the horizon: transport and reduced $CO_2$ emissions in the UK by 2030', *Transport Policy*, **4**, 377–87.
Jackson, C., A. Smith and M. Conner (2003), 'Applying an extended version of the Theory of Planned Behaviour to physical activity', *Journal of Sports Sciences*, **21**, 119–33.
Kitamura, R. (1988), 'An evaluation of activity-based travel analysis', *Transportation*, **15**, 9–34.
Klöckner, C.A. and A. Blöbaum (2010), 'A comprehensive action determination model: toward a broader understanding of ecological behaviour using the example of travel mode choice', *Journal of Environmental Psychology*, **30**(4), 574–86.
Koppelman, F.S. and P.K. Lyon (1981), 'Attitudinal analsis of work/school travel', *Transportation Science*, **15**(3), 233–54.
Kurani, K. and M.E.H. Lee-Gosselin (1996), *Synthesis of Past Activity Analysis Applications*, 2–5 June, Texas Transportation Institute.
McFadden, D. (2001), 'Disaggregate behavioural travel demand's RUM side – a 30 years retrospective', in D.A. Hensher (ed.), *Travel Behaviour Research: The Leading Edge*, Oxford: Pergamon Press, pp. 17–64.
McNally, M.G. (2000), 'The activity based approach', in D.A. Hensher and K.J. Button (eds), *Handbook of Transport Modelling*, Amsterdam: Pergamon, pp. 53–70.
McNally, M.G. and C. Rindt (2008), *The Activity-based Approach*, Center for Activity Systems Analysis, Institute of Transportation Studies (UCI), University of California Irvine.
Metcalfe, R. and P. Dolan (2012), 'Behavioural economics and its implications for transport', *Journal of Transport Geography*, **24**, September, 503–11.
Ogilvie, D., F. Bull, J. Powell et al. (2011), 'An applied ecological framework for evaluating infrastructure to promote walking and cycling: the iConnect study', *American Journal of Public Health*, **101**(3), 473–81.
Price, T.J. and S.D. Probert (1995), 'Role of road transport in UK's energy policy', *Applied Energy*, **50**, 1–22.
Prillwitz, J. and S. Barr (2011), 'Moving towards sustainability? Mobility styles, attitudes and individual travel behaviour', *Journal of Transport Geography*, **19**(6), 1590–600.
Ritsema van Eck, J., G. Burghouwt and M. Dijst (2005), 'Lifestyles, spatial configurations and quality of life in daily travel: an explorative simulation study', *Journal of Transport Geography*, **13**, 123–34.
Salomon, I. and M. Ben-Akiva (1983), 'The use of the life-style concept in travel demand models', *Environment and Planning A*, **15**, 623–38.
Schwanen, T. and K. Lucas (2011), 'Understanding auto motives', in K. Lucas, E. Blumenberg and R. Weinberger (eds), *Auto Motives: Understanding Car Use Behaviours*, Bingley, UK: Emerald Group Publishing.
Schwanen, T. and P.L. Mokhtarian (2007), 'Attitudes toward travel and land use and choice of residential neighborhood type: evidence from the San Francisco Bay area', *Housing Policy Debate*, **18**(1), 171–207.

Schwanen, T., D. Banister and J. Anable (2012), 'Rethinking habits and their role in behaviour change: the case of low-carbon mobility', *Journal of Transport Geography*, **24**, September, 522–32.

Schwartz, S.H. (1977), 'Normative influence on altruism', in L. Berkowitz (ed.), *Advances in Experimental Psychology*, New York: Academic Press, pp. 221–79.

Stern, P. (2000), 'Toward a coherent theory of environmentally significant behavior', *Journal of Social Issues*, **56**(3), 407–24.

Stern, P., T. Dietz, T. Abel, G.A. Guagnano and L. Kalof (1999), 'A value-belief-norm theory of support for social movements: the case of environmentalism', *Human Ecology Review*, **6**, 81–97.

Swait, J. (2001), 'A non-compensatory choice model incorporating attribute cutoffs', *Transportation Research Part B: Methodological*, **35**(10), 903–28.

Thøgersen, J. (2009), 'Promoting public transport as a subscription service: effects of a free month travel card', *Transport Policy*, **16**, 335–43.

Triandis, H.C. (1977), *Interpersonal Behavior*, Monterey, CA: Brooks/Cole Publishing.

Verplanken, B., I. Walker, A. Davis and M. Jurasek (2008), 'Context change and travel mode choice: combining the habit discontinuity and self-activation hypotheses', *Journal of Environmental Psychology*, **28**(2), 121–7.

Weinberger, R. and F. Goetzke (2011), 'Drivers of auto ownership: the role of past experience and peer pressure', in K. Lucas, E. Blumenberg and R. Weinberger (eds), *Auto Motives: Understanding Car Use Behaviours*, Bingley, UK: Emerald Group Publishing.

# 7. Technology

**Justin Bishop**

## INTRODUCTION

Carbon dioxide equivalent greenhouse gas emissions arise when fuel with embodied carbon is used by vehicles to satisfy their propulsion requirements. The emission of other air pollutants arises from the use of such fuel. Mass is the largest determinant of the energy and power required by a vehicle to complete a journey (Goldberg, 2000). Additional energy is required to overcome both the drag[1] (forces), which acts on the vehicle in motion, and the losses from the conversion of potential energy to kinetic energy, such as in an engine.

The total demand for transport accounted for 27 per cent of global energy use in 2008. Road transport was responsible for 74 per cent of that total, followed by marine, aviation and rail at 16 per cent, 11 per cent and 2 per cent, respectively (IEA, 2010, see also Chapter 8). There are two components that determine the extent to which energy use and emissions of greenhouse gases and other pollutants, more importantly, may be reduced or eliminated from transport. First, novel technologies can yield an immediate, but one-off, efficiency boost when incorporated either with new vehicles or as retrofits to those already in service. Second, energy use and associated emissions can be reduced on an ongoing basis through more efficient vehicle operation. This chapter provides an overview of both options across air, sea and land transport modes and through to 2050. A third option to reduce emissions by fuel switching is discussed briefly in this chapter, and in more detail in Chapter 8.

With respect to new technologies, the focus is on improving onboard energy conversion to propulsion and decreasing both drag and vehicle mass. Regulations, business practices, public acceptance and physical infrastructure constraints influence the extent to which vehicles may be operated more efficiently, and these aspects are also considered in this chapter. Demand side measures, such as decreasing the distance travelled or the amount of transport service required, may provide the largest potential to reduce absolute energy use and emissions. Such measures are beyond the scope of this chapter and are not covered (see Part II of this book).

In general, emissions from transport are normalized to both the distance travelled and service provided, such as $gCO_2$ per passenger-km and $gCO_2$ per freight tonne-km. Normalized efficiency metrics reflect the efficiency of the transport service, but mask the absolute changes in emissions, distance travelled and amount of service supplied. For example, from 1971 to 1998, the energy intensity of the US airlines' fleets was reduced by 60 per cent, but demand increased by 300 per cent offsetting any emission reduction benefits (Lee et al., 2001).

## AIR TRAVEL

The energy used to transport one passenger one kilometre decreased by 66 per cent from 1960[2] to 2007[3]. From 1959 to 1995 the majority of the decrease in the energy used per passenger-km was attributed to engine efficiency improvements, aerodynamic improvements and higher load factors (Morrell, 2009; Lee, 2010).

Jet engines (or turbofans) are used on 84 per cent of aircraft above 9000 kg maximum take-off mass (ICAO, 2009). Geared turbofans and unducted turbofans using contra-rotating blades offer the best propulsion efficiency improvements by 2020. Fuel burn can be reduced by 12 per cent using a geared link (geared turbofan). The central shaft of a conventional turbofan engine must rotate at high speeds to compress efficiently the intake air. Inserting a gearbox allows the fan to turn at a lower speed than the central shaft. Therefore, a larger fan may be used that increases the bypass ratio (Lawrence, 2009). Fuel burn decreases as the bypass ratio – ratio of the mass of air bypassing the engine core to the mass of air passing through the engine core – increases. However, there is a trade-off between the improvement in propulsion efficiency with increasing bypass ratio and both the mass and aerodynamic penalties associated with larger engine nacelles (Parker, 2009). Unducted turbofans with contra-rotating blades can be 10–30 per cent more efficient than a ducted turbofan at acceptable noise levels, but require a marginal reduction in aircraft operating speed (Green, 2009; Parker, 2009; IATA, 2010a; McCollum et al., 2010; Szodruch et al., 2011).

Beyond 2020, conventional turbofan efficiency may be improved by 20 per cent by increasing the temperatures and pressures in the combustor, while complying with regulation on nitrous oxide ($NO_x$) levels (Parker, 2009). Exhaust gas may be recovered via an intercooled recuperative core to preheat air entering the combustor. Introducing preheated, pressurized air to the combustor can reduce the work necessary by the compressor and associated fuel burn by 5–10 per cent. New materials with a tolerance of high temperatures, such as single crystal nickel alloys, may be used in engines to accommodate increased turbofan peak temperature (Parker, 2009).

After 2030, superconducting motors and generators may be used to decouple the mechanical link between the combustion and propulsion processes. Such machines have a high efficiency and a power per unit mass that is comparable to conventional turbines. However, introducing superconducting machines requires auxiliary apparatus to provide low temperature cooling. Substituting electrical cables for the mechanical link permits greater flexibility in both where the engines are placed in the airframe (packaging) and in their operation and is consistent with the aim of achieving an all-electric aircraft (Luongo et al., 2009). Conceptually, small nuclear reactors and heat exchangers may be used in place of conventional combustors to generate a heated air stream (Mitrica et al., 2010).

Aerodynamic drag can be reduced both by preserving laminar flow along the aircraft surface and by minimizing the effects of an already turbulent flow. Currently, winglets appended to the wing tips of aircraft reduce fuel burn by 7 per cent (IATA, 2009; McCollum et al., 2010). By 2020, nano-materials in advanced coatings (Parker, 2009) and adjustable control surfaces may reduce fuel burn by 1–5 per cent (IATA, 2009). Improving aircraft shape can maintain natural laminar flow far beyond the rear of the fuselage to reduce fuel burn by 5–15 per cent (Green, 2009; IATA, 2009; Szodruch et al., 2011).

By 2030, the use of airframes that change both shape and properties to reduce aerodynamic drag (morphing) will reduce fuel burn by 5–10 per cent (IATA, 2009). Hybrid and full laminar flow control systems may exist to suck the air flow close to the fuselage skin (Green, 2009) to achieve fuel burn reductions of 10–15 per cent (IATA, 2009) and 20 per cent, respectively. The greatest advance in reducing aerodynamic drag will result from integrating propulsion and lift to form the blended wing body or flying wing (Figure 7.1), reducing fuel burn by 15–32 per cent (Green, 2009; McCollum et al., 2010). Incorporating full laminar flow control with the blended wing body could reduce fuel burn a further 10–55 per cent (Green, 2009; McCollum et al., 2010).

The trend of airframe material composition is moving towards 65 per cent composites and 15 per cent aluminium (Lawrence, 2009). Mass savings of 15 per cent may be achieved by substituting traditional aluminium with composite materials (Green, 2009). By 2020, additional mass savings of 1–3 per cent will be made by replacing the pneumatic and hydraulic lines in conventional powertrains with electrical (fly-by-

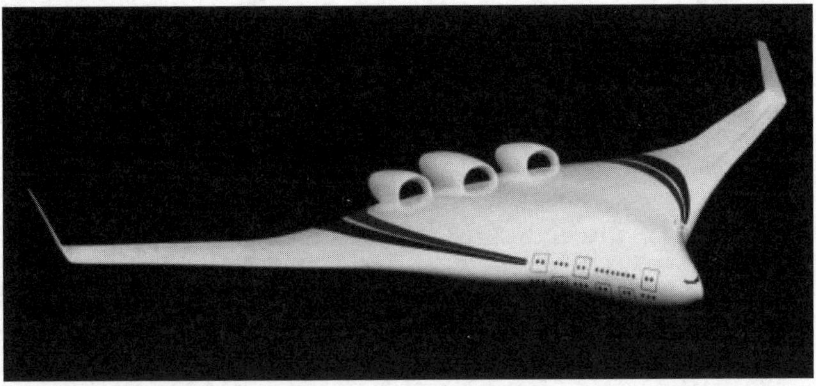

*Figure 7.1 NASA Blended wing body X-48 aircraft*

wire, Parker, 2009) or optical (fly-by-light, IATA, 2009) connections. In general, every 1 per cent mass saved leads to a 0.25–0.75 per cent fuel burn reduction (Lawrence, 2009).

The most promising technological solution to reduce emissions from aircraft is the blended wing body airframe incorporating an optimally placed and more efficient propulsion system. The service lifetime of aircraft precludes rapid fleet turnover and delays the efficiency boost that novel technologies can provide. Moreover, the blended wing body with integrated propulsion is a radical shift from conventional aircraft design. Therefore, a number of years may be required to develop the confidence needed to deploy the technology, establish regulatory frameworks and secure public acceptance.

Meanwhile, improving the operating efficiency of aircraft may reduce emissions from the existing fleet in the near to medium term. Aircraft operate inefficiently outside their design cruise mode, that is, when not cruising. More efficient air traffic control measures and ground operations to reduce the amount of flying outside the cruise mode may reduce emissions by 8–18 per cent (Green, 2009; Lawrence, 2009; McCollum et al., 2010; Szodruch et al., 2011). Aircraft can use less energy in flight by: flying in formation, similar to birds; cruising at an optimum (and lower) altitude to reduce the impacts of nitrous oxides and contrails; cruising at a reduced speed, which has impacts on wing mass and thrust required; refuelling by air-to-air transfer to reduce the number of refuelling take-offs and landings; engaging in continuous descent approach; and better scheduling of aircraft departure and arrival to avoid 'holding' in the air until clearance for landing is granted. While on the ground, aircraft can use less energy, employing mains electricity (shore power) while on stand and either single-engine taxiing or 'engineless' taxiing by using an electric ground tug. Some innovations, such as formation flying, air-to-air refuelling and ground towing, represent a major departure from how the industry operates currently. Therefore, their implementation will be limited by both the time and costs associated with a change from business as usual.

## MARITIME TRANSPORT

Emissions generated from transporting one tonne of freight one kilometre decreased by a median factor of three across the main ship types – general cargo, container, bulk and tankers – from 1900 to 1995 (IMO, 2009). Many of the technologies to improve the efficiency of transport services provided by ships already exist in other industries. Therefore, the introduction of novel technologies to reduce emissions from maritime transport will be based on cost, the regulatory environment and the timetable for fleet renewal. Conventional ship engines operate at low and medium speeds and utilize either heavy fuel oil or marine distillate fuel oil to satisfy propulsion and auxiliary loads. The next generation of ships may benefit from switching to gas turbines that operate on the Brayton cycle. These turbines are more efficient and offer better potential for exhaust gas recovery.

In the future the mechanical link between generation and propulsion may be decoupled by using superconducting electric motors (Kalsi et al., 2006). The electricity to supply these motors may be generated by Brayton cycle[4] turbines or from the direct chemical processes in modular, reliable fuel cells. Fuel cells may be used alone or hybridized with gas turbines to boost system efficiency (McCollum et al., 2010). Solid oxide fuel cells are best suited to satisfy the heavy-duty needs of a ship. Their high operating temperature can be used both to convert liquid fossil fuels to hydrogen in a combined cycle with a gas turbine to deliver combined heat, power and steam to propulsion and non-propulsion (also referred to as auxiliary, hotel or service) loads (Sun et al., 2010).

Solar photovoltaics and wind turbines have low conversion efficiency and their output is dependent on both the weather and the time of day. The limited deck area constrains the power output from these technologies to a fraction of the non-propulsion loads. Kites and sails can offset main engine use and fuel consumption: notably, the best designed ship travelling at low speed in the most favourable wind conditions can demonstrate fuel burn saving of 44 per cent (IMO, 2009). The complexity of harnessing power from waves precludes its use on ships.

Co-optimizing the hull and propellers may yield 5–20 per cent fuel burn savings in still water (IMO, 2009). Hydrodynamic drag on a ship is a function of weather, the ship configuration and both the shape and the condition of its hull.[5] Friction drag may be decreased by improving hull surface quality and applying hull coatings to reduce fouling (Eide et al., 2009; Cariou, 2011). Washing the hull to remove fouling may reduce fuel burn by 5 per cent (IMO, 2009). The drag along propeller surfaces can be reduced by 5 per cent by cleaning/polishing the propeller surfaces and by using a smaller number of blades (surfaces).

Propulsion efficiency increases by 5–10 per cent for large diameter propellers rotating at low angular velocity (IMO, 2009). The efficiency of transferring energy to the water is dependent on the condition of its flow around the propeller. Pre-swirl stators condition the water on approach to the propellers, yielding 3–8 per cent fuel burn reductions. Ducting focuses the fluid flow through the propellers, offering 5–20 per cent fuel burn savings. Contra-rotating propellers improve the transfer of power to the water, reducing fuel burn by 3–6 per cent. Finally, post-swirl devices, such as free-rotating vane wheels (Grim wheels) and an asymmetric hull shape (better integration of the propellers and rudders) can reduce fuel burn by 10 per cent and 1–9 per cent, respectively (Corbett et al., 2009; Eide et al., 2009; IMO, 2009).

Mass can be reduced through substituting high strength steels for conventional ones. Extensive use of composites is currently reserved for high performance craft. Reducing ship mass reduces the area of the hull in contact with the water (wetted surface area) for all loads (IMO, 2009).

The most promising technological options for reducing emissions from maritime transport are improvements in propulsion efficiency and the co-optimization of the hull and propeller to better deliver the engine power to the water. Ships have long service lives in the same way as aircraft. Therefore, in the near to medium term emissions from ships may be reduced best through increasing the efficiency of the voyage. Energy use by ships in motion is dependent on the voyage characteristics. Voyage optimization can be introduced fleet-wide to reduce emissions by 5–50 per cent (Eide et al., 2009; IMO, 2009; McCollum et al., 2010) and comprises: weather routing to exploit favourable winds and sea currents; reduced and constant cruising speed; optimum trim (difference between how deep the bow of the ship sits in the water relative to the stern) to balance load and ballast; minimum directional changes; and use of larger ships to reduce the number of trips. Just-in-time arrival can reduce emissions by 1–5 per cent. Small speed reductions reduce fuel burn while maintaining engine operation near its most efficient point. Engine derating – the operation of an engine at less than its rated maximum power – to prolong its life permits further slow steaming by better matching the engine's most efficient operating point to demand (IMO, 2009). Many ships are currently operated by non-owner parties that do not have an economic incentive to improve voyage efficiency. Other financial incentives, such as demand for fast product delivery, crew costs and berth timetables, may reduce the feasibility of reduced voyage speeds.

## LAND TRANSPORT

Rail, light-duty and heavy goods vehicles satisfy passenger and freight transport demands over land. Light-duty passenger vehicles[6] comprise the largest portion of energy use and emissions from transport globally at 45 per cent, followed by freight trucks at 25 per cent, buses, 6 per cent and rail, 2 per cent (Ribeiro et al., 2007, see also Chapter 8).

### Light-duty and Heavy-duty Vehicles

Many of the technology development options for both light-duty and heavy goods vehicles are mature enough to satisfy the requirements of mass production or have been demonstrated on an experimental basis. The efficiency of the conventional (port-injected spark ignition) petrol engine may be improved best by switching to direct injection compression ignition, derating and incorporating exhaust gas recovery. Direct injection engines use compression of the fuel-air mixture to start combustion. Conversely, a spark is used to start the combustion of the fuel-air mixture in conventional petrol engines. The switch to direct injection engines may lead to a 40 per cent increase in efficiency (Cole et al., 1999; Greene and Plotkin, 2001; Kutlar et al., 2005; IEA, 2008; Sharpe, 2009). Downsizing (or derating) the engine to better match average, rather than maximum, loads, combined with exhaust gas recovery can yield 2–20 per cent energy efficiency improvements. Variable compression ratios and variable valve lift and timing can each contribute 10 per cent efficiency improvements. The transmission gearbox is used to match the range of wheel speeds to the range of engine speeds. Optimum gear selection seeks to satisfy desired wheel speed and highest efficiency engine operation. More gears allow optimum engine operation for a wider range of wheel speeds, yielding a 4–10 per cent decrease in fuel consumption. By switching off the engine when idle fuel burn can be reduced by 0.5–3 per cent (Leduc et al., 2003; Kutlar et al., 2005; IEA, 2008, 2009; DOE, 2009; Sharpe, 2009; NRC, 2010). Engine start/stop is particularly useful for buses and heavy goods vehicles operating in urban environments and can raise fuel savings potential to 50 per cent (Greszler, 2009).

The greatest potential to reduce fuel burn in both light-duty and heavy goods vehicles lies in the ability to recover kinetic energy while braking and store it onboard for use during later accelerations. Typically, an internal combustion engine is combined with onboard energy storage to form a hybridized powertrain for both light-duty and heavy goods vehicles. Light-duty vehicle powertrain hybridization can reduce fuel burn by 15-33 per cent (Weiss et al., 2000; Ahluwalia et al., 2004; Auer et al., 2006; IEA, 2008, 2009; Sharpe, 2009; Boretti, 2010), up to 55 per cent with a plug-in feature to facilitate recharging of the energy store from mains power (IEA, 2009). Hybrid powertrains in heavy goods vehicles may use pressurized fluid (hydraulics) to store energy. Using the hydraulics to accelerate the vehicle from rest (launch assist) may reduce emissions by 36 per cent on duty cycles with frequent stops (Odhams et al., 2010).

In light-duty vehicles kinetic energy may be recovered using the compact and efficient permanent magnet motor that can satisfy the vehicle duty cycle without a transmission (Husain, 2003; Williamson et al., 2007; Boretti, 2010). The recovered energy may be stored: chemically in either nickel metal hydride or lithium ion batteries; electrically in ultracapacitors; or mechanically in flywheels. The real-world operation of a road vehicle with regenerative braking results in frequent and large flows of power to and from the wheels. Batteries are better suited to store energy than to provide power. Conversely, ultracapacitors and flywheels satisfy power demands more efficiently than they store energy. In heavy goods vehicles hydraulic regenerative braking in a parallel configuration is more space efficient than the equivalently rated electric and kinetic regenerative systems. However, energy storage in batteries may be used to satisfy non-propulsion demands, such as the sleeper cab hotel loads, and to eliminate engine idling when stationary, reducing fuel burn by 10 per cent (Midgley and Cebon, 2010).

In the future the internal combustion engine may be replaced with a fuel cell. Proton exchange membrane fuel cells have low operating temperatures, good dynamic response and are the most appropriate for light-duty vehicles. Solid oxide fuel cells are suitable for heavy goods vehicles with predictable duty cycles and where non-propulsion loads may be served well by the high temperature exhaust.

Improving light-duty vehicle aerodynamics involves trade-offs to satisfy various market demands, such as low wind noise, luggage capacity, rear seating headroom and aesthetics (Hucho and Sovran, 1993). Therefore, the ability for energy use to be reduced significantly through drag reduction is limited. Pressure drag is largely determined by the shape of the vehicle in general and of its rear end in particular (Hucho and Sovran, 1993). Fuel burn reductions of up to 8 per cent can be achieved by: rounding the front (leading) edge of the vehicle; covering the chassis underside; using vortex generators at the rear (trailing) edge of the roof to delay the separation of air flow as it passes over the rear of the car; using body coatings to reduce friction drag; redesigning the engine and passenger compartment ventilation; and reducing the frontal area (Beaudoin et al., 2004; Koike et al., 2004; IEA, 2008, 2009; Kobayashi et al., 2009; Odhams et al., 2010). In heavy goods vehicles fuel burn may be reduced by 25 per cent by: changing the tractor cab-behind-engine to a cab-over-engine configuration to eliminate the pressure drag discontinuities associated with the horizontal bonnet; sealing the tractor-trailer gap; and modifying the trailer to include leading-edge fairings; underbody smoothing; side skirts; and vortex stabilizers ('boat tails') (Ogburn et al., 2008; Greszler, 2009).

One-third of the mechanical energy at the wheels is used to overcome rolling resistance (TRB, 2006), which is a function of tyre size, material properties, the road conditions for which it was designed, inflation pressure and vehicle loading. Reducing rolling resistance can improve fuel efficiency by up to 6 per cent (TRB, 2006; Ogburn et al., 2008; IEA, 2008, 2009; Greszler, 2009; Odhams et al., 2010). Combining aerodynamic and tyre rolling resistance improvements can reduce the propulsion load by 40 per cent in heavy goods vehicles (Ogburn et al., 2008).

Vehicle mass may be reduced by material substitution and component redesign (Bjelkengren, 2008). Saving 10 per cent of primary mass reduces fuel burn by 5–7 per cent (Cheah, 2010). However, 'mass begets mass' such that secondary savings of 1.04 kg can proceed from every 1 kg primary savings (Bjelkengren, 2008). Reducing the mass of an unladen heavy goods vehicle is attractive as more cargo can be transported for a fixed gross vehicle mass. In heavy goods vehicles additional mass saving options include using derated engines and replacing tandem tyres with wide-based 'super singles' to reduce fuel burn by 3 per cent (Greszler, 2009).

Hybridization and electrification of vehicle powertrains present the greatest technological innovations to reducing the emissions from both light-duty and heavy goods vehicles. This technology is present already in some light-duty and heavy goods vehicles in service. Owners and operators, however, may be either unwilling or unable to bear the higher capital costs that accompany vehicles with hybridized/electrified powertrains. Yet, legislation exists now to compel manufacturers to deliver high efficiency vehicles, many of which will incorporate such technologies. Therefore, new ownership models may be required to bridge the gap between high vehicle cost and affordability/consumer willingness to pay.

Eco-driving of both light-duty and heavy goods vehicles may reduce fuel burn by 5–20 per cent (Ribeiro et al., 2007) and comprises optimum gear selection, gentle acceleration and decelerations and reduced vehicle speed

(Greszler, 2009). For heavy goods vehicles, logistic options to reduce fuel burn include scheduling deliveries when roads are less congested (arguing against night-time curfews) to reduce the energy used per trip and increasing load factors (to more than 50 per cent using 'back-loads') to reduce the total number of trips. Trailer volume can be increased by 5 per cent by lowering the floor by 15 cm to allow more cargo per trip. In the limit, additional trailers may be appended to a tractor to create a long combination vehicle. Such vehicles use up to 18 per cent less fuel and can maintain their manoeuvrability in urban settings when the trailers are fitted with steering axles (Ogburn et al., 2008; Odhams et al., 2010). Drivers may need training and feedback from in-vehicle indicators to realize the full potential of eco-driving. Changing driver attitude to one of minimizing fuel consumption and emissions may be the largest barrier to overcome. This is difficult particularly for heavy goods vehicles operating under competitive business practices, such as just-in-time delivery schemes and short delivery times (see Chapter 10). Ultimately, external factors, such as road congestion, may eliminate many of the technological and operational improvements of road vehicles described above.

**Rail Transport**

Many train locomotives are powered by diesel engines or electric motors. The emissions from locomotives driven with electric motors can be less than those from diesel engines, depending on the carbon intensity of the electricity generation (Givoni et al., 2009). The efficiency of both of these technologies can be improved with the ability to recover kinetic energy and store it onboard (Hillmansen and Roberts, 2007; UIC, 2011). Ultracapacitors are seen as the most suitable technology for auxiliary power supply in trains. Other onboard energy storage media include compressed air, magnetic fields and flywheels (Lu et al., 2008).

Most conventional high-speed trains employ efficient and robust electric traction motors (Garcia, 2010) supplied by overhead lines. The use of alternating current motors avoids the losses incurred when mains alternating current is converted to direct current. In the future fossil fuels may be converted in an onboard reformer to hydrogen for use in a fuel cell to provide electricity to traction motors (Meegahawatte et al., 2010).

Over 75 per cent of the energy used by a train locomotive is to overcome aerodynamic drag at high speeds. Bogie interference is responsible for 47 per cent of the drag created, while skin friction and overhead equipment contribute 30 per cent and 20 per cent, respectively (Schetz, 2001). Optimizing the exterior dimensions of a train can reduce drag significantly. For example, for the same seating capacity, a double-deck, wide-bodied

*Source:* Photograph taken by user JakeLM under the Creative Commons Attribution 2.5 Generic license.

*Figure 7.2 Shanghai Maglev train on the track in 2005*

train suffers from 40 per cent less drag than a single deck, normal body train (Garcia, 2010). Additional improvements can be made by optimizing the shape of the lead nose, streamlining bogies and using telescoping pantographs (the hinged electric-rod devices that conduct electric current from overhead lines) complete with fairings (Schetz, 2001; Garcia, 2010). Drag on intermodal freight trains may be reduced by moving the cargo as close to the train head as possible and minimizing the space between cargo containers (slot utilization) (Lai et al., 2008).

Material substitution is an option for reducing the mass of trains. All trains can benefit from removing empty carriages from the end to reduce total vehicle mass, allowing optimum cruise speed (Lai et al., 2008). Uniquely for electric train locomotives, any recovered energy from braking events may be retransmitted to the wider public network. Such a process may reduce or eliminate onboard energy storage and the associated mass penalty (Garcia, 2010).

In the limit, the physical coupling of the train wheels to the track can be replaced with magnetic interactions (Figure 7.2). A combination of magnetic repulsion (electrodynamic suspension) for levitation, magnetic attraction for guidance and linear synchronous motors for propulsion (Lee et al., 2006) allows such magnetic levitation trains to operate at greater than 500 km per hour. At these speeds magnetic levitation trains compete favourably with air travel for distances greater than 800 km. Magnetic levitation is the greatest technological innovation for rail. This high cost technology may see widespread deployment only in the long term and be reserved for passenger transport, where a premium is paid for speed of travel (analogous to aircraft).

During train operation energy use can be reduced by 15 per cent by: using intelligent air refreshing/conditioning based on passenger occupancy, not capacity; driving economically, including coasting; and increasing speed on downhill grades (Bai et al., 2009; Garcia, 2010). Changes to infrastructure to reduce track gradient would assist in reducing energy use by trains during operation, but there is significant lock-in to the existing rail infrastructure. Therefore, there are large cost and time penalties associated with modifying the infrastructure to reduce track gradients, increase the radii of corners for higher travel speeds and satisfy the demand for higher capacity.

## FUELS

In addition to technological developments and changing operation practices, emissions can be reduced by changing the fuel used (see Chapter 8). However, changing the energy sources is also dependent on technological development. The fuels in use by the majority of vehicles currently in service are: kerosene by aircraft; heavy fuel oil or marine distillate fuel oil by ships; petrol and diesel by light-duty vehicles; diesel by heavy goods vehicles; and diesel or electricity by trains. Switching from carbon-intensive fossil fuels to alternatives can break the link between fuel use and emissions. Novel fuel options include: naphtha; natural gas; liquefied petroleum gas; ethanol; biodiesel; synthetic diesel; dimethyl ether; methanol; and hydrogen. The well-to-tank energy use (and associated emissions) can be significantly greater than the fossil fuel incumbents. Therefore, reducing vehicle emissions by using novel fuels is dependent on identifying supply pathways with low greenhouse gas emissions. In general, the pathways with the least embodied energy and emissions may use: renewable sources; the waste or by-products of primary resources either directly or as part of the production process; or employ end-of-pipe processes to reduce or capture emissions.

Novel transport fuels must overcome the challenges of infrastructure lock-in, cost and yield. Conventional and new liquid fuels are produced centrally and may be supplied efficiently along existing infrastructure at low cost. Conversely, centrally produced gaseous fuels suffer from supply/distribution infrastructure lock-out. Onsite production may overcome the infrastructure challenge, but at both a higher cost and greater process inefficiency. Finally, the yield of fuels using plant feedstocks can be significantly lower than those from fossil fuels. Therefore, it is unlikely that large quantities of conventional and novel fuels from fossil fuel resources will be displaced if growing demand for transport is to be met.

Aircraft in service can use hydroprocessed renewable jet fuel and synthetic fuels formed via the Fischer-Tropsch process as direct, drop-in replacements (IATA, 2010b). The main Fischer-Tropsch feedstocks are algae, camelina and jatropha for aircraft (IATA, 2010b). By 2020, the emissions from new aircraft may be reduced by at least 15 per cent through more extensive use of liquefied petroleum gas, liquefied natural gas, compressed natural gas, ethanol, blended butanol and biofuels (IATA, 2009). Further into the future, the zero greenhouse gas emissions from hydrogen during vehicle operation may be balanced by the contribution of water vapour to the creation of contrails. Liquefied natural gas is an alternative fuel for ships, reducing greenhouse emissions by 15 per cent (Eide et al, 2009; IMO, 2009). At present, using nuclear power for ship propulsion is limited to military vessels (McCollum et al., 2010). Light-duty and heavy goods vehicles can use synthetic diesels as direct, drop-in replacements for diesel. Compressed natural gas, liquefied petroleum gas, hydrogen and bioethanol can be used with existing light-duty and heavy goods vehicle engine technology. Dimethyl ether and methanol are the alternative fuel options with the largest potential to meet the volume of demand by heavy goods vehicles. Biofuels have the potential to reduce operational emissions from ships and land vehicles by at least 75 per cent (Greszler, 2009). The use of liquid fossil fuels (or their novel replacements) in solid oxide fuel cells for ship, train and heavy goods vehicle applications allows hydrogen to be produced on demand and mitigates the challenges of storage.

## CONCLUSION

The main ways to reduce fuel burn by vehicles in motion using technology is to increase the efficiency of converting fuel to propulsion and to reduce both vehicle mass and the magnitude of forces opposing motion. Emissions may be decoupled from the energy used to satisfy propulsion needs by using novel fuels with low environmental impacts.

The most promising technologies for reducing energy use by future aircraft include geared turbofans, unducted turbofans, a blended wing body structure with full laminar flow control and a fly-by-light control scheme. Energy efficient ships of the future will include advanced and integrated hull-propeller design paired to either gas turbines with exhaust gas recovery or solid oxide fuel cells with superconducting motors. Efficient light-duty and heavy goods vehicles of the future will combine onboard energy storage with an efficient, derated internal combustion engine as part of hybridized powertrains. In the limit, light-duty vehicle powertrains may become all-electric and be refuelled from mains electricity by plugging in, or hybridized with a fuel cell or both. High-speed and energy efficient trains will continue to be driven by electric traction motors and transmit any recovered kinetic energy from braking to the grid. Magnetic levitating trains will realize the limits of high-speed mass transport on land.

All vehicles can benefit from reducing their mass and drag. Reducing vehicle mass delivers compounding benefits and is accomplished through material substitution and component design. Decreasing the drag on vehicles in motion is accomplished often by: reducing the skin roughness; delaying the separation of air flow by optimized design of the vehicle shape; and covering the discontinuities in vehicle shape, such as between tractor and trailer or between train bogies.

Alternative fuels are available to meet the specific performance needs of all transport modes: hydroprocessed renewable jet and Fischer-Tropsch fuels are drop-in replacements for kerosene in aircraft; liquefied natural gas in turbines to replace heavy fuel oils in low-speed marine diesel engines on ships; and synthetic diesels, dimethyl ether and methanol to replace the diesel used in light-duty and heavy goods vehicles; and electricity for trains. Novel transport fuels must overcome a number of challenges to successfully displace the incumbents. Such challenges include reduction of the embodied carbon in their supply from well to tank, distribution infrastructure lock-out, primary resource yield and the final cost of fuel delivered to the filling station.

The challenge of reducing the greenhouse gas emissions from transport can be overcome by addressing both the technology in vehicles and the way in which the vehicles are operated. Novel technologies provide a one-off efficiency increase in the energy used in vehicles. Better vehicle operation can yield continuous, large reductions in energy use and be introduced to vehicles both in service now and in the future. Eco-driving, including reducing travel speed, can offer immediate benefits to all modes of transport. However, vehicle operations are limited by physical infrastructure constraints, regulations and long-term industry practices. Therefore, some of the large benefits from improving vehicle operation may be realized only in the long term alongside radical technological innovations and fleet turnover. Furthermore, transport costs may fall on account of improved energy efficiency, through improvements to technology and/or operations. Consequently, the absolute volume of transport services demanded may increase in a rebound effect and lead to an increase in overall emissions.

## ACKNOWLEDGEMENT

The author acknowledges the funding provided for this work by the Oxford Martin School.

## NOTES

1. Drag comprises a friction force – acting along the vehicle surface – and a pressure, which acts normal to the vehicle surface.
2. See Boeing 707-120B specifications online at http://www.boeing.com/commercial/airports/acaps/707.pdf (accessed 20 September 2012).
3. See Airbus A380-800 specifications online at http://www.airbus.com/aircraftfamilies/passengeraircraft/a380family/a380-800/specifications/ (accessed 20 September 2012).
4. The Brayton cycle consists of a compressor, combustion chamber and turbine. Fresh air is drawn into the compressor. The pressurized air is mixed with the fuel in the combustion chamber and ignited. The high temperature combustion products drive a turbine attached to a drive shaft. Some of the turbine energy is used to drive the compressor (Ramgir and Sable, 2006).
5. Air resistance is small in comparison to other drag forces, though side winds acting on the ship superstructure can affect the drift and voyage efficiency (IMO, 2009).
6. Cars including, for example, minivans and sport utility vehicles (SUVs).

## REFERENCES

Ahluwalia, R.K., X. Wang, A. Rousseau and R. Kumar (2004), 'Fuel economy of hydrogen fuel cell vehicles', *Journal of Power Sources*, **130**(1–2), 192–201, doi 10.1016/j.jpowsour.2003.12.061.

Auer, J., G. Sartorelli and J. Miller (2006), 'A gatekeeper energy management strategy for ECVT hybrid vehicle propulsion utilising ultracapacitors', Paper presented at the IET Hybrid Vehicle Conference, Coventry, UK, pp. 79–90.

Bai, Y., B. Mao, F. Zhou, Y. Ding and C. Dong (2009), 'Energy-efficient driving strategy for freight trains based on power consumption analysis', *Journal of Transportation Systems Engineering and Information Technology*, **9**(3), June, 43–50.

Beaudoin, J.F., O. Cadot, J.L. Aider et al. (2004), 'Cavitation as a complementary tool for automotive aerodynamics', *Experiments in Fluids*, **37**(5), 763–8, doi 10.1007/s00348-004-0879-y.

Bjelkengren, C. (2008), 'The impact of mass decompounding on assessing the value of vehicle lightweighting', PhD thesis, Massachusetts Institute of Technology, Massachusetts, available at http://msl.mit.edu/theses/Bjelkengren_C-thesis.pdf (accessed 20 September 2012).

Boretti, A. (2010), 'Comparison of fuel economies of high efficiency diesel and hydrogen engines powering a compact car with a flywheel based kinetic energy recovery system', *International Journal of Hydrogen Energy*, **35**(16), 8417–24, doi 10.1016/j.ijhydene.2010.05.031.

Cariou, P. (2011), 'Is slow steaming a sustainable means of reducing $CO_2$ emissions from container shipping?', *Transportation Research Part D: Transport and Environment*, **16**(3), 260–4, doi 10.1016/j.trd.2010.12.005.

Cheah, L.W. (2010), 'Cars on a diet: the material and energy impacts of passenger vehicle weight reduction in the U.S.', PhD thesis, Massachusetts Institute of Technology, Massachusetts, available at http://web.mit.edu/sloan-autolab/research/beforeh2/files/LCheah PhD thesis 2010.pdf (accessed 20 September 2012).

Cole, R.L., R.B. Poola and R. Sekar (1999), 'Emissions and fuel economy of a vehicle with a spark-ignition, direct-injection engine: Mitsubishi Legnum GDI™', ANL/ESD/TM-149, Argonne National Laboratory.

Corbett, J.J., H. Wang and J.J. Winebrake (2009), 'The effectiveness and costs of speed reductions on emissions from international shipping', *Transportation Research Part D: Transport and Environment*, **14**(8), 593–8, doi 10.1016/j.trd.2009.08.005.

DOE (2009), *FY2009 Progress Report for Advanced Combustion Engine Research and Development*, Washington, DC: Office of Vehicle Technology, US Department of Energy.

Eide, M.S., Ø. Endresen, R. Skjong, T. Longva and S. Alvik (2009), 'Cost-effectiveness assessment of $CO_2$ reducing measures in shipping', *Maritime Policy and Management*, **36**(4), 367–84, doi 10.1080/03088830903057031.

Garcia, A. (2010), *High Speed, Energy Consumption and Emissions*, Paris: International Union of Railways.

Givoni, M., C. Brand and P. Watkiss (2009), 'Are railways "climate friendly"?', *Built Environment*, **35**(1), 70–86.

Goldberg, L.H. (2000), *Green Electronics, Green Bottom Line: Environmentally Responsible Engineering*, Woburn, MA: Butterworth-Heinemann, Chapter 24.

Green, J.E. (2009), 'The potential for reducing the impact of aviation on climate', *Technology Analysis and Strategic Management*, **21**(1), 39–59, doi 10.1080/09537320802557269.

Greene, D.L. and S.E. Plotkin (2001), 'Energy futures for the US transport sector', *Energy Policy*, **29**(14), 1255–70, doi 10.1016/S0301-4215(01)00071-4.

Greszler, A. (2009), *Reducing Climate Impacts in the Transportation Sector*, New York: Springer, doi 10.1007/978-1-4020-6979-6.

Hillmansen, S. and C. Roberts (2007), 'Energy storage devices in hybrid railway vehicles: a kinematic analysis', *Proceedings of the Institution of Mechanical Engineers, Part F: Journal of Rail and Rapid Transit*, **221**(1), 135–43, doi 10.1243/09544097JRRT99.

Hucho, W.H. and G. Sovran (1993), 'Aerodynamics of road vehicles', *Annual Review of Fluid Mechanics*, **25**, 485–537, doi 10.1146/annurev.fl.25.010193.002413.

Husain, I. (2003), *Electric and Hybrid Vehicles*, CRC Press.

IATA (2009), *The IATA Technology Roadmap Report*, 3rd edn, Montreal: International Air Transport Association, available at https://www.escholar.man

chester.ac.uk/api/datastream?publicationPid=uk-ac-man-scw:106699&datastreamId=FULL-TEXT.PDF (accessed 20 September 2012).

IATA (2010a), *Airlines International Environment Supplement*, Montreal: International Air Transport Association, available at http://www.iata.org/pressroom/airlinesinternational/april-2010/Documents/airlines-international-supp-environment-042010.pdf (accessed 20 September 2012).

IATA (2010b), *IATA 2010 Report on Alternative Fuels*, 5th edn, Montreal: International Air Transport Association, available at http://www.iata.org/ps/publications/Documents/IATA%202010%20Report%20on%20Alternative%20Fuels.pdf (accessed 20 September 2012).

ICAO (2009), *Annual Report of the Council*, Doc. 9921, Montreal: International Civil Aviation Organization, available at http://www.icao.int/publications/Documents/9921_en.pdf (accessed 20 September 2012).

IEA (2008), *Energy Technology Perspectives 2008 – Scenarios & Strategies to 2050*, Paris: International Energy Agency.

IEA (2009), *Transport, Energy and $CO_2$*, International Energy Agency, available at http://www.iea.org/media/weowebsite/factsheets/fact_sheets_WEO_2009.pdf (accessed 20 September 2012).

IEA (2010), *International Energy Agency (IEA) World Energy Balances*, ESDS International (Mimas), University of Manchester.

IMO (2009), *Second IMO GHG Study 2009*, London: International Maritime Organization, available at http://www.imo.org/blast/blastDataHelper.asp?data_id=27795&filename=GHGStudyFINAL.pdf (accessed 20 September 2012).

Kalsi, S.S., B.B. Gamble, G. Snitchler and S.O. Ige (2006), 'The status of HTS ship propulsion motor developments', Paper presented at the IEEE Power Engineering Society (PES) General Meeting, Montreal, pp. 1–5, doi 10.1109/PES.2006.1709643.

Kobayashi, S., S. Plotkin and S.K. Ribeiro (2009), 'Energy efficiency technologies for road vehicles', *Energy Efficiency*, **2**(2), 125–37, doi 10.1007/s12053-008-9037-3.

Koike, M., T. Nagayoshi and N. Hamamoto (2004), 'Research on aerodynamic drag reduction by vortex generators', *Mitsubishi Motors Technical Review*, **16**, 11–16, available at http://www.mitsubishi-motors.com/corporate/about_us/technology/review/e/pdf/2004/16E_03.pdf (accessed 20 September 2012).

Kutlar, O.A., H. Arslan and A.T. Calik (2005), 'Methods to improve efficiency of four stroke, spark ignition engines at part load', *Energy Conversion and Management*, **46**(20), 3202–20, doi 10.1016/j.enconman.2005.03.008.

Lai, Y.-C., C.P.L. Barkan and H. Onal (2008), 'Optimizing the aerodynamic efficiency of intermodal freight trains', *Transportation Research Part E: Logistics and Transportation Review*, **44**(5), 820–34, doi 10.1016/j.tre.2007.05.011.

Lawrence, P. (2009), 'Meeting the challenge of aviation emissions: an aircraft industry perspective', *Technology Analysis and Strategic Management*, **21**(1), 79–92, doi 10.1080/09537320802557327.

Leduc, P., B. Dubar, A. Ranini and G. Monnier (2003), 'Downsizing of gasoline engine: an efficient way to reduce $CO_2$ emissions', *Oil and Gas Science and Technology*, **58**(1), 115–27, doi 10.2516/ogst:2003008.

Lee, H.W., K.C. Kim and J. Lee (2006), 'Review of maglev train technologies', *IEEE Transactions on Magnetics*, **42**(7), 1917–25, doi 10.1109/TMAG.2006.875842.

Lee, J.J. (2010), 'Can we accelerate the improvement of energy efficiency in aircraft systems?', *Energy Conversion and Management*, **51**(1), 189–96, doi 10.1016/j.enconman.2009.09.011.

Lee, J.J., S.P. Lukachko, I.A. Waitz and A. Schafer (2001), 'Historical and future trends in aircraft performance, cost, and emissions', *Annual Review of Energy and Environment*, **26**, 167–200.

Lu, S., D.H. Meegahawatte, S. Guo, S. Hillmansen, C. Roberts and C.J. Goodman (2008), 'Analysis of energy storage devices in hybrid railway vehicles', Paper presented at the International Conference on Railway Engineering (ICRE), Hong Kong, pp. 1–6.

Luongo, C.A., P.J. Masson, T. Nam et al. (2009), 'Next generation more-electric aircraft: a potential application for HTS superconductors', *IEEE Transactions on Applied Superconductivity*, **19**(3), 1055–68, doi 10.1109/TASC.2009.2019021.

McCollum, D.L., G. Gould and D.L. Greene (2010), 'Greenhouse gas emissions from aviation and marine transportation: mitigation potential and policies', UCD-ITS-RP-10-01, Institute of Transportation Studies, University of California at Davis, available at http://escholarship.org/uc/item/5nz642qb (accessed 20 September 2012).

Meegahawatte, D., S. Hillmansen, C. Roberts, M. Falco, A. McGordon and P. Jennings (2010), 'Analysis of a fuel cell hybrid commuter railway vehicle', *Journal of Power Sources*, **195**(23), 7829–37, doi 10.1016/j.jpowsour.2010.02.025.

Midgley, W.J.B. and D. Cebon (2010), *Comparison of Regenerative Braking Technologies for Heavy Vehicles*, CUED/C-MECH/TR.96, Cambridge: Department of Engineering, University of Cambridge.

Mitrica, B., V. Stanciu, M. Petre, M.O. Dima, C. Petre and I. Precup (2010), 'Contributions regarding the aircraft nuclear propulsion', in *American Institute of Physics Conference Proceedings*, 7th International Conference of the Balkan Physical Union, Alexandroupolis, Greece, vol. 1203, pp. 23–8, doi 10.1063/1.3322435.

Morrell, P. (2009), 'The potential for European aviation $CO_2$ emissions reduction through the use of larger jet aircraft', *Journal of Air Transport Management*, **15**(4), 151–7, doi 10.1016/j.jairtraman.2008.09.021.

NRC (2010), *Technologies and Approaches to Reducing the Fuel Consumption of Medium- and Heavy-duty Vehicles*, Washington, DC: National Academies Press.

Odhams, A.M.C., R.L. Roebuck, Y.J. Lee, S.W. Hunt and D. Cebon (2010), 'Factors influencing the energy consumption of road freight transport', *Proceedings of the Institution of Mechanical Engineers, Part C: Journal of Mechanical Engineering Science*, **224**(9),1995–2010, doi 10.1243/09544062JMES 2004.

Ogburn, M., L. Ramroth and A.B. Lovins (2008), 'Transformational trucks: determining the energy efficiency limits of a class-8 tractor-trailer', Rocky Mountain Institute, available at http://old.rmi.org/images/PDFs/Transportation/RMITransformationalTruck Study 080709compressed (accessed 16 May 2011).

Parker, R. (2009), 'From blue skies to green skies: engine technology to reduce the climate-change impacts of aviation', *Technology Analysis and Strategic Management*, **21**(1), 61–78, doi 10.1080/09537320802557301.

Ramgir, M.S. and M.J. Sable (2006), *Gas Turbines and Jet Propulsion*, 1st edn, Pune: Technical Publications Pune.

Ribeiro, S.K., S. Kobayashi, M. Beuthe et al. (2007), 'Transport and its infrastructure', in *Climate Change 2007: Mitigation. Contribution of Working Group III to the Fourth Assessment Report of the Intergovernmental Panel on Climate Change*, Cambridge: Cambridge University Press.

Schetz, J.A. (2001), 'Aerodynamics of high speed trains', *Annual Review of Fluid Mechanics*, **33**, 371–414, doi 10.1146/annurev.fluid.33.1.371.

Sharpe, R.B.A. (2009), 'Technical options for fossil fuel based road transport', European Commission Directorate-General Environment and AEA Technology plc, available at http://www.eutransportghg2050.eu/cms/assets/EU-Transport-GHG-2050-Paper-1-Technical-options-for-ffuel-road-transport-11-02-10-FINAL.pdf (accessed 20 September 2012).

Sun, J., J. Steve and C. Kennell (2010), 'Feasibility and design implications of fuel cell power for sealift ships', *Naval Engineers Journal*, **122**(3), 87–102, doi 10.1111/j.1559-3584.2010.00274.x.

Szodruch, J., W. Grimme, F. Blumrich and R. Schmid (2011), 'Next generation single aisle aircraft – requirements and technological solutions', *Journal of Air Transport Management*, **17**(1), 33–9, doi 10.1016/j.jairtraman.2010.10.007.

TRB (2006), *Tires and Passenger Vehicle Fuel Economy: Informing Consumers, Improving Performance – Special Report 286*, National Academy of Sciences, USA, available at http://onlinepubs.trb.org/onlinepubs/sr/sr286.pdf (accessed 20 September 2012).

UIC (2011), *Rail and Sustainable Development*, Paris: International Union of Railways.

Weiss, M.A., J.B. Heywood, E.M. Drake, A. Schafer and F.F. Au Yeung (2000), 'On the road in 2020: a life-cycle analysis of new automobile technologies', Technical Report MIT EL 00-003, Massachusetts Institute of Technology, available at http://web.mit.edu/energylab/www/pubs/el00-003.pdf (accessed 20 September 2012).

Williamson, S.S., A. Emadi and K. Rajashekara (2007), 'Comprehensive efficiency modeling of electric traction motor drives for hybrid electric vehicle propulsion applications', *IEEE Transactions on Vehicular Technology*, **56**(4),1651–572, doi 10.1109/TVT.2007.896967.

# 8. Future energy mix and transport

**Martino Tran**

## INTRODUCTION

This chapter analyses future energy mix options for a more sustainable transport system. The chapter takes an integrated approach by assessing key issues between energy use and transport and different supply side opportunities for decarbonizing transport. Key issues behind increasing energy use in transport include economic activity, population growth, urban migration, demographics and increasing demand in emerging economies. These drivers will influence future energy demand and transport systems. The energy mix is then assessed, covering conventional oil, gas and nuclear as well as important renewable options and implications surrounding future electricity supply. The chapter closes by highlighting the link between transport and energy security, prompting the need for fuel diversification in light of climate policy.

## ENERGY AND TRANSPORT

### Economic Activity

Economic activity is the primary driver of demand for all energy services. Energy demand generally grows in line with gross domestic product (GDP), though typically at a lower rate. There can also be some decoupling with increasing energy efficiency, combined with a transition from heavy industry to a more service-based economy among industrialized countries. Between 1980 and 2008 world primary energy demand increased on average 0.59 per cent per year for every percentage point of GDP growth (in real purchasing power parity). This gross income elasticity of demand has fluctuated over time, falling from 0.64 in the 1980s to 0.46 in the 1990s but rebounding to 0.67 over 2000–08, largely due to expansion of energy-intensive manufacturing in China. Generally, income elasticity of demand tends to be higher for countries at an early stage of economic development than for mature economies, where saturation effects curb income-driven increases in demand (IEA, 2010).

Between 2005 and 2050 global GDP is projected to grow four-fold, reaching US$227 trillion per year. In Europe and Japan GDP is projected to nearly double while North America's GDP is projected to increase by 2 to 2.5 times its current level. The main growth will be in transition economies and developing countries. China and India's GDP will grow nearly ten-fold. Chinese GDP is projected to be 70 per cent higher than that of the US while Indian GDP will reach nearly the same level as OECD Europe. The global share of GDP for the G8+5[1] countries is projected to increase from a current 69 per cent to 73 per cent by 2050 (IEA, 2008). An important concern is that as purchasing power increases among emerging economies per capita vehicle ownership will increase, if those countries follow the same development path as the Organisation for Economic Co-operation and Development (OECD). With increasing wealth among emerging economies there is an opportunity to leap-frog straight to clean technology and pursue a portfolio of public transport options (rail, bus, tram and so on), and the potential for avoiding heavy investment into fossil fuel-based transport technology and infrastructure.

**Population Growth**

Population growth is a key driver of the amount and type of energy used. This in turn has direct implications for an increase in aggregate transport carbon emissions. The world's population is now 6.7 billion. From 1980 to 2008 the population expanded by 1.5 per cent per year. In the medium term projections expect slower growth consistent with long-term historical trends, from 1.1 per cent per year from 2008–20 and 0.7 per cent from 2020–35 (UNPD, 2009). The OECD population is expected to increase by 0.4 per cent per year on average over 2008–35. It is predicted that the non-OECD population will expand by 1 per cent per year, from 5.5 billion in 2008 to 7.2 billion in 2035, reaching 85 per cent of the global total. Russia is the only major non-OECD country with a predicted fall from 142 million in 2008 to 126 million in 2035. Africa is expected to grow the fastest, averaging a population increase of 1.9 per cent per year between 2008 and 2035 while for non-OECD Asia it is expected to increase from 3.5 billion to 4.3 billion (IEA, 2010). Over the period to 2035, the population of India is expected to overtake that of China, reaching 1.47 billion in 2035. Over the period to 2050, it is expected that the world population will increase by over 37 per cent from 6.5 billion to 9.2 billion led by non-OECD countries (UN, 2007; IEA, 2010). The G8+5 population is expected to drop from 56 per cent to 48 per cent of the total by 2050 (IEA, 2008).

**Urban Migration**

The increase in world population will occur in urban areas with rural populations declining in most regions except in Africa. Rapid urbanization will increase demand for modern energy services, which are more available in towns and cities. Urban dwellers have increased by 10 per cent over the last 25 years and are projected to increase to 60 per cent by 2030 and 70 per cent by 2050 (UN, 2005). In 2009 the urban population surpassed the rural population for the first time in history. In 2008 the urban population was 3.3 billion and is projected to reach 5.2 billion by 2035, mostly in non-OECD countries (UNPD, 2010). Between 2000 and 2030 urban populations in Asia are projected to increase from 1.36 billion to 2.64 billion, in Africa from 294 million to 742 million, and in Latin America and the Caribbean from 394 million to 609 million. Consequently, developing countries will have 80 per cent of the world's urban population by 2030. Africa and Asia will have almost seven out of every ten urban inhabitants in the world (UNFPA, 2007).

Urbanization coupled with changes in demographics and population growth will affect future energy use and therefore carbon dioxide ($CO_2$) emissions. Currently, the global median age is 28 years, increasing to 38 years by 2050. The proportion of the population that is 60 years or over is projected to rise from 11 per cent in 2007 to 22 per cent in 2050 (UN, 2007). This ageing will affect energy consumption since lifestyles and needs of older people differ from those of younger people. A recent study by O'Neil et al. (2010) shows that slowing population growth could provide 16–29 per cent of the suggested $CO_2$ emissions reductions by 2050 to avoid dangerous climate change. This is based on global population scenarios ranging from +1.7 to –1.5 billion relative to a midpoint of ~9 billion in 2050. Rapid urbanization also presents an opportunity for diffusion of low carbon transport, where short-range electric vehicles could meet the functionality and performance expected by consumers. This will depend on the development of charging infrastructure, where there is greater potential to mitigate range anxiety through battery swap stations or public charging within cities rather than over longer distances. For more discussion on urbanization and transport see Chapter 5.

**Emerging Economies**

Demand for automobile travel is projected to increase more than three-fold, with freight growing even faster. Efficiency gains of 30–50 per cent from conventional petrol-based vehicle technologies will be insufficient to outweigh demand growth (IEA, 2008). Aggregate growth in global transport energy demand is projected to average 1.3 per cent per year from 2008 to 2035. Transport's share of total final energy consumption remains about 27 per cent towards 2035 (IEA, 2008). All of the growth in transport demand comes from non-OECD regions and inter-regional bunkers, with transport energy demand declining in the OECD. Although biofuels and electricity for plug-in hybrid electric vehicles (PHEVs) and full battery electric vehicles (BEVs) will take an increasing market share from road transport fuels, oil-based fuels will continue to dominate transport energy demand (IEA, 2010).

Regions and countries show different patterns in both energy use per capita and the types of transport fuel used. North America (except Mexico) averaged over 1200 kilotonnes oil equivalent (ktoe) per person in 2006, while Africa averaged less than 100 ktoe per person. This shows the differences both in the amount of travel undertaken and in the fuels used for that travel. The relative efficiency of vehicles is a much smaller factor in overall energy use. Within the transport sector light-duty vehicles (LDVs[2]) and trucks account for about 75 per cent of travel-related energy used in the OECD and about 68 per cent in non-OECD countries. Aircraft and marine shipping account for most of the remainder (IEA, 2009).

Due to a 94 per cent reliance on oil, transport is the second largest source of $CO_2$ emissions at 6.3 gigatonnes (Gt) in 2008 or 24 per cent of total $CO_2$ emissions, compared to power generation (40 per cent), industry (16 per cent), buildings (12 per cent) and agriculture and non-energy use (8 per cent) (IEA, 2010). Two main factors influence transport $CO_2$ emissions: change in total volume of travel and fuel efficiency of the travel mode. Given the relatively low average rates of vehicle ownership in emerging economies coupled with rising GDP growth rates, vehicle travel is expected to increase. Consequently, $CO_2$ emission reductions made in industrialized countries will be offset by increasing energy and transport demand in emerging economies.

Worldwide rates of passenger LDVs and two- and three-wheeler ownership in 2005 show that two- and three-wheelers outnumber passenger LDVs in Asia. In industrialized countries the rate of growth in passenger LDV ownership has declined in recent years, possibly due to decreasing population growth. But in developing countries rates of car ownership are growing rapidly. The emergence of low-cost cars due to economies of scale in developing countries may further accelerate LDV ownership rates. The number of motorized two-wheelers is also projected to grow rapidly (IEA, 2009).

Due to rising incomes, LDV ownership rates have rapidly increased in emerging economies. In 2005 LDV ownership rate (vehicles per 1000 people) in China was 11 compared to 424 and 559 in OECD Europe and

North America, respectively. From 2000 to 2005 total sales of LDVs in China increased from 700000 to 3.1 million, a 340 per cent increase, and will continue to rise (IEA, 2009). Over the same period North American LDV sales increased 3 per cent from 16.9 to 17.5 million, while OECD European sales decreased 2 per cent from 16.7 to 16.3 million (IEA, 2009).

## ENERGY MIX

### Oil

In 2008 the transport sector was 94 per cent dependent on oil, consuming 2150 million tonnes oil equivalent (Mtoe) and is expected to increase on average 1.4 per cent per year reaching 3102 Mtoe by 2035 (IEA, 2010). Most of the oil consumed worldwide is used in transport with nearly all recent growth in oil use due to growth in transport. In 2006 more than 60 per cent of petroleum products used in OECD countries and 50 per cent in non-OECD countries were used as transportation fuels. Energy diversification within the transport sector has been a high priority for many oil-importing countries but very few countries, other than Brazil, have made progress. There are a number of reasons for transport dependence on oil related to the characteristics of oil and oil products, including high energy densities, they are easy to handle and transport, an extensive oil-based vehicle stock and infrastructure, and historically low market prices.

Peak oil has increasingly become a major policy issue and potentially an important impetus for diversification into renewable energy. Although there are around 70000 oil fields in the world, only about 25 fields account for 25 per cent of global crude oil production, 100 fields account for 50 per cent of total production and up to 500 fields account for two-thirds of cumulative discoveries. Most of these 'giant' fields are relatively old and past peak production. Most of the rest will begin to decline within the next ten years with few new giant fields expected to be found. The average rate of decline from fields that are past their peak of production is 6.5 per cent per year globally, while the corresponding rate of decline from all current producing fields is 4 per cent per year. This implies that around 3 mb/d (million barrels per day) of new capacity must be added each year to maintain current production levels, which is equivalent to a new Saudi Arabia coming on stream every three years. Decline rates are on an upward trend as more giant fields enter decline, as production shifts towards smaller, younger and offshore fields and as changing production methods lead to more rapid post-peak decline. As a result, more than 68 per cent of current crude oil production capacity may need to be replaced by 2030, simply to prevent production from falling (UKERC, 2009; Owen et al., 2010).

The ultimately recoverable resources (URR) of a region depend upon economic and technical factors as much as geology and can only be estimated when exploration is advanced. The timing of a global peak (or plateau) in conventional oil production may be estimated to within decadal accuracy assuming a particular value for the global URR and no significant disruptions to the oil market. The timing of the global peak for conventional oil production is relatively insensitive to assumptions about the size of the global resource. For a wide range of assumptions about the global URR of conventional oil and the shape of the future production cycle, the date of peak production is estimated to lie between 2009 and 2031. Although this range appears wide, in the face of a peak, it is a narrow window of lead time to develop substitute fuels (UKERC, 2009). Given the nearly complete dependence of transport on oil, the potential fallout from a global peak highlights the need to diversify transport fuels (Owen et al., 2010).

### Gas

Gas demand is set to expand rapidly in two emerging sectors: (1) feedstock for gas-to liquids (GTL) plants and (2) road transport fuel. The recent fall in the price of gas relative to oil, especially in North America, has stimulated interest in using natural gas as a road transport fuel. Today, natural gas vehicles are common in only a few countries and the global use of compressed natural gas (CNG) as a road fuel is marginal. The biggest potential lies with heavy-duty vehicles (trucks and buses), as the costs of installing refuelling infrastructure for light-duty vehicles and adapting cars to run on gas are likely to limit the growth of CNG use for passenger vehicles (Lopez et al., 2009). There is scope for increased CNG consumption in non-OECD Asia and Latin America, but the greatest potential could be in North America, where extensive supplies of unconventional gas may hold gas prices down, making CNG an attractive alternative to diesel for heavy-duty vehicles. The International Energy Agency (IEA, 2010) projects that North American gas use for road transport could potentially increase from 0.9 billion cubic metres (bcm) in 2008 to 12 bcm by 2035, and global use could rise from 18 bcm to 61 bcm over the same period (IEA, 2010).

Remaining resources of natural gas are abundant compared to oil, and are sufficient to meet the projected increase in global demand. The largest uncertainty for supply over the next 25 years is whether sufficient and timely investment will be made in developing gas resources and how much exploitation will cost. Proven reserves of gas have increased steadily since the 1970s, as reserve additions have outpaced production by a wide margin. Proven reserves were 184 trillion cubic metres (tcm) at the end of 2008, almost twice as high as 20 years ago and equivalent to 58 years of production at current rates (IEA, 2010).

Currently, Russia, Iran and Qatar hold 54 per cent of the world's proven gas reserves. The OECD holds 10 per cent of the total. It is estimated that the remaining recoverable resources of conventional gas are 404 tcm. There is also an estimated 380 tcm of unconventional recoverable gas, increasing total recoverable gas resources to nearly 800 tcm, or 250 years of current production (IEA, 2010). Global gas demand scenarios suggest cumulative investment along the gas supply chain of $7.1 trillion (2009 US dollars), equalling $270 billion per year from the present to 2035 (IEA, 2011). Approximately two-thirds of capital spending of $175 billion per year is needed upstream for new greenfield projects and due to a decline at existing fields. The majority of this investment is needed in non-OECD countries.

**Nuclear**

Nuclear power generation has the capacity to provide large-scale, virtually $CO_2$-free, electricity. However, safety is still an important concern, highlighted by recent events such as in Fukashima, Japan. How much safety and proliferation concerns influence longer term nuclear development for various countries remains to be seen. The Intergovernmental Panel on Climate Change (IPCC, 2007) suggests that nuclear power could reach 18 per cent of total electricity supply by 2030. Yet, there is uncertainty due to outstanding issues surrounding safety, weapons proliferation and waste disposal that could constrain further growth. Nuclear power could contribute 6 per cent of $CO_2$ savings based on the construction of 30 gigawatts (GW) of capacity each year between the present and 2050 (IEA, 2008). Much of this would occur in China and India.

A global scenario suggests that nuclear power generation will play an important role with capacity increasing from 368 GW to 570 GW in 2050 (IEA, 2008). As most of the standing capacity must be replaced in the next 45 years, the baseline scenario implies on average more than ten new reactors per year. Without this capacity replacement, more $CO_2$-emitting capacity would need to be built and emissions would be even higher. Modelling work has further shown that further nuclear expansion to 2000 GW could be cost-effective, but largely at the expense of fossil-fuelled plants with carbon capture and sequestration (CCS). However, it remains unclear whether such an increase would be acceptable or feasible, due to fuel reprocessing on a massive scale posing a challenge for non-proliferation. The main insight from global projections is that an expansion of nuclear power generation does not substantially reduce emissions because nuclear growth would occur largely at the expense of fossil-fuelled power plants with CCS.

**Renewables**

In 2008 renewable energy (RE) accounted for 12.9 per cent of the 492 Exajoules (EJ) of total primary energy supply (Figure 8.1). The largest contributor was biomass (10.2 per cent) with about 60 per cent of traditional biomass used in cooking and heating applications in developing countries. Hydropower was 2.3 per cent, and all other RE sources 0.4 per cent. RE contributed about 19 per cent of the global electricity supply (16 per cent hydropower, 3 per cent other RE), and biofuels contributed 2 per cent of the global road transport fuel supply. Traditional biomass (17 per cent), which is made of fuel wood, agricultural and forest residues and solid waste often used in cook stoves, more sustainable modern biomass (8 per cent), which uses those materials for electricity and heat production, and solar thermal and geothermal energy (2 per cent) together fuelled 27 per cent of the total global demand for heat (IPCC, 2011).

Deployment of RE has increased rapidly in recent years (Figure 8.2). Government policies, declining cost of RE technologies, changes in fossil fuel prices and increases in energy demand have accelerated deployment of RE. Between 2008–09 the cumulative installed capacity of RE increased: wind power by 32 per cent (38 GW), hydropower by 3 per cent (31 GW), grid-connected photovoltaic by 53 per cent (7.5 GW), geothermal power by 4 per cent (0.4 GW) and solar hot water/heating by 21 per cent (31 GW). Biofuels increased to 3 per cent in 2009 from 2 per cent in 2008. The annual production of ethanol increased to 1.6 EJ (76 billion litres) by the end of 2009 and biodiesel to 0.6 EJ (17 billion litres) (IPCC, 2011). Of the approximately 300 GW of new electricity generating capacity added globally from 2008 to 2009, 140 GW came from RE. Developing countries account for 53 per cent of the global RE electricity generation capacity (IPCC, 2011).

Currently, biofuels, hydrogen ($H_2$) and electricity are widely regarded as key renewable forms of energy competing for application in the transport sector. However, this depends on whether or not they are derived from renewable sources.

**Biofuels**

Global production of biofuels was 52 Mtoe in 2009. The US and Brazil account for roughly 75 per cent of global biofuel production. Ethanol

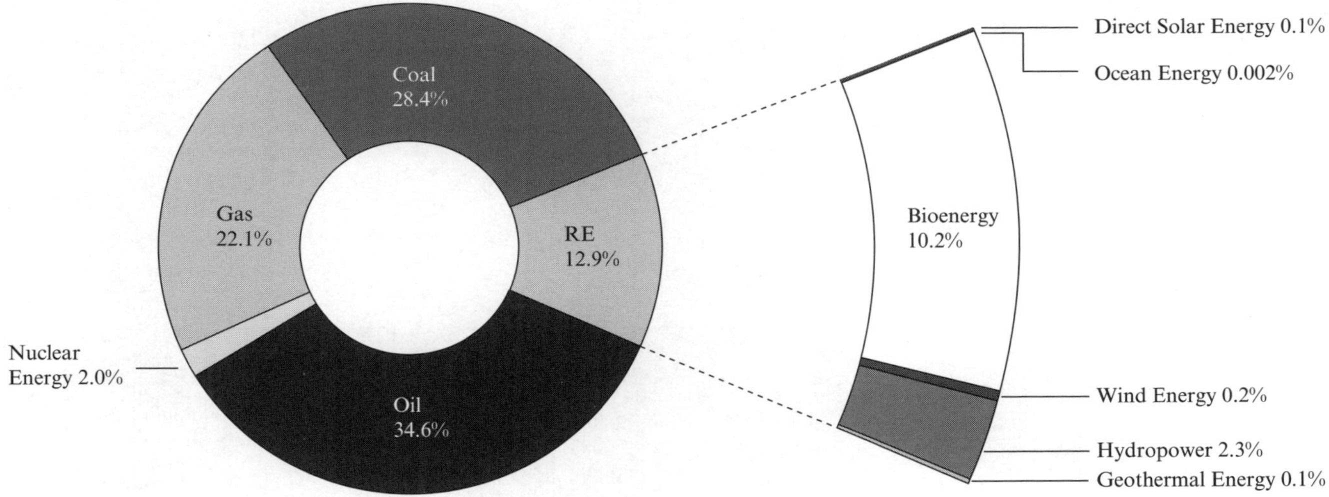


*Source:* IPCC (2011).

*Figure 8.1 Shares of energy sources in total global primary energy supply in 2008 (492 EJ)*

Primary Solid Biomass for Heat and Electricity Applications
Hydropower
Global Primary Energy Supply (EJ per year)
0
10
20
30
40
50
60
Biofuels (incl. Biogas)
Wind Energy
Geothermal Energy
Solar Thermal Energy
Municipal Solid Waste (Renewable Share)
Global Primary Energy Supply (EJ per year)
0
1
2
3
4
5

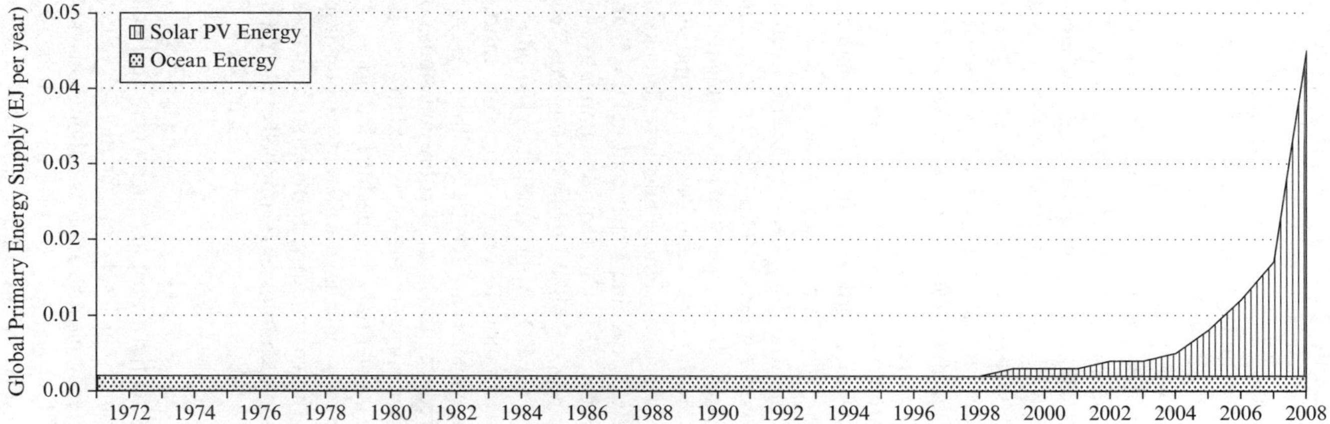


*Source:* IPCC (2011).

*Figure 8.2 Historical development of global primary energy supply from renewable energy from 1971 to 2008*

accounts for about 75 per cent of global production of biofuels for transport. Investment in biofuels was severely affected by the economic and financial crisis in 2008/2009, falling by over 60 per cent compared to 2008 as a result of lower oil prices and a drop in demand for transport fuels, but is projected to rebound in the short term (IEA, 2010).

Despite rapid growth over the past decade, biofuels accounted for only 3 per cent of global road transport fuel demand in 2009. Policy support helped to increase production in the US by nearly 30 per cent per year on average from 2002 to 2009, overtaking Brazil as the largest producer in 2005. In 2009, of the 3 per cent biofuel share used in road transport fuel, Brazil had the highest share of biofuels in its road transport fuel mix at 20 per cent. Currently, biofuels are used almost exclusively for road transport, but interest in biofuels for aviation is growing (IEA, 2010).

Biofuel demand is expected to increase due to rising oil prices and government support, prompted by energy security and environmental concerns. Biofuels receive more government support than any other RE source or carrier. In 2009 government support was $20 billion with the highest levels in the US and the European Union (EU). Ethanol receives most of this at $13 billion. Between 2010 and 2020 government support is expected to be on average $45 billion per year, increasing to $65 billion per year from 2021 to 2035, with 60 per cent for ethanol and 40 per cent for biodiesel. Advanced biofuels using lignocellulosic feedstocks[3] are assumed to enter the market by 2020, reaching 36 per cent in OECD and 5 per cent in non-OECD countries by 2035. This requires cumulative investment in the production capacity of $335 billion from 2010 to 2035 with about 60 per cent occurring in the OECD. Approximately $180 billion is for conventional production of ethanol, 10 per cent for conventional biodiesel and the remainder for advanced biofuels (IEA, 2010). Although biofuels have the potential to reduce transport carbon emissions, there can be negative implications when agricultural land is used to grow biofuel crops, rather than food supplies, potentially increasing food prices. Additionally, competing biofuel pathways will have different carbon impacts, where bioelectricity has been shown to outperform ethanol across a range of feedstocks and conversion technologies (Campbell et al., 2009).

**Hydrogen**

Hydrogen is another potential transport fuel where fuel cells use $H_2$ to generate electricity onboard vehicles. Hydrogen fuel cell vehicles (HFCVs) benefit from high efficiency electric motors relying on compressed $H_2$ tanks to store energy. Hydrogen can be produced on board the vehicle from ethanol or other liquid fuels containing $H_2$ but it is expected that it will be more cost-effective to store $H_2$ directly onboard vehicles after producing it elsewhere. Of the 8 EJ (about 190 Mtoe) of $H_2$ currently produced each year, 40 per cent is used in chemical processes, 40 per cent in refineries and 20 per cent in other areas. In 2003 48 per cent of all $H_2$ was produced from natural gas, 30 per cent from oil and off-gases from refineries and chemical plants, 18 per cent from coal and 4 per cent from electrolysis (IEA, 2009).

There are currently significant barriers for the market diffusion of HFCVs, including high vehicle costs, lack of charging and production infrastructure, the more disruptive nature of the technology and hydrogen costs, which will likely be higher than the cost of electricity. Moreover, the large up-front investment for a distribution network heightens the risk of failure in case actual HFCV development does not meet industry and policy expectations, creating an additional barrier for development. Global investment costs for hydrogen infrastructure are estimated to be in trillions of US dollars. Given the advantages of internal combustion engine vehicles (ICEs) and existing infrastructure, without policy interventions and financial support from governments around the world, market forces alone will not deliver a system shift from hydrocarbon to hydrogen (Service, 2004).

Currently, there are market failures that do not account for externalities across competing drive-trains. When externality costs for oil supply security, air pollution and greenhouse gas (GHG) emissions are calculated for the full fuel cycle, BEVs have lower costs than current ICEs and HFCVs have the lowest externality costs of any option including advanced gasoline ICEs, hybrid electric vehicles (HEVs), diesel, Fischer–Tropsch liquids and compressed natural gas (Ogden et al., 2004). For health benefits alone, it is estimated that HFCVs, where hydrogen is produced from wind power, could save 2300 to 4000 lives per year and $32–180 billion per year in the US compared to hybrids (Jacobsson et al., 2005). Policies are needed to lower upfront costs and level the playing field with ICEs.

## Electricity

Most regions in the world currently do not generate sufficient low $CO_2$ electricity to capture the $CO_2$ emissions reduction potential of full BEVs and PHEVs (IEA, 2009). Optimistic scenarios show the global electric vehicle market share reaching 90 per cent in 2050 and electricity demand reaching 580 Mtoe, or 17 per cent of world electricity demand, requiring some 2000 GW of additional capacity. In that scenario 4 Gt of $CO_2$ are abated from the transport sector by 2050. But the key assumption here is that the global average electricity carbon intensity must drop below 100

$gCO_2 kWh^{-1}$ by 2050 from an average 460–550 $gCO_2 kWh^{-1}$ in 2005 (IEA, 2010).

In 2006 average electricity emissions ranged from a low of 190 $gCO_2 kWh^{-1}$ in Latin America to a high of 944 $gCO_2 kWh^{-1}$ in India. One of the largest potential electric vehicle markets is China but the current grid mix is 788 $gCO_2 kWh^{-1}$. Even across industrialized countries the differences can be significant. In the UK a PHEV operating on a current grid mix of 450 $gCO_2 kWh^{-1}$ emits 110 $gCO_2 km^{-1}$. The same PHEV in the US would emit 183 $gCO_2 km^{-1}$, or 60 per cent more than its UK counterpart due to a higher grid mix of 670 $gCO_2 kWh^{-1}$ (King, 2007).

When looking across power trains within the US, under current electricity generation profiles PHEV carbon performance is only slightly better than a HEV. And if carbon-intensive generation further expands PHEVs will become more carbon intensive than HEVs and will only be slightly better than a conventional petrol vehicle assuming the phase-out of coal and greater reliance on imported liquefied natural gas (LNG), which has higher GHG emissions than the domestic supply. If PHEVS become dependent on imported LNG for electricity this could have negative implications for both GHG emissions and energy security. Moreover, if commercialization of PHEVs is to be matched with low carbon electricity generation, investment into decarbonized power must be made now to be ready for increasing PHEV market penetration over the next 20 to 30 years (Samaras and Meisterling, 2008).

## Smart Grids

A smart grid is an electricity network that uses digital and other advanced technologies to monitor and manage the transport of electricity from all generation sources to meet the varying electricity demands of end-users. This includes electricity transmission and distribution systems interfacing with generation, storage and end-users. Smart grids potentially have an enabling role for other low carbon energy technologies, including electric vehicles, variable RE sources and demand response. Smart grids are an evolving set of technologies that will be deployed at different rates in a variety of settings around the world, depending on local commercial attractiveness, compatibility with existing technologies, regulatory developments and investment frameworks (DOE, 2009).

Electricity is the fastest-growing component of total global energy demand, with consumption expected to increase over 150 per cent from 2007 to 2050 (IEA, 2011). In OECD countries, where modest growth rates are based on high levels of current demand, smart grid technologies can potentially reduce transmission and distribution losses, while optimizing the use of existing infrastructure. In developing regions with high growth there is the potential to avoid infrastructural lock-in and leap-frog to smart grid technologies offering more efficient operation. It is expected that smart grid technologies could help reduce demand by providing consumers with the information they need to use less energy or use it more efficiently. Unfortunately, in many regions the necessary technology investment is hindered by existing regulatory structures, which often have long approval processes and do not capture the benefits of new innovative technologies, combined with ineffective business models that have not attracted larger participation (Boots et al., 2010).

The world's electricity systems face a number of challenges, including ageing infrastructure, continued growth in demand, rapid integration of variable RE sources, electric vehicles, security of supply and lower carbon emissions. These challenges are inter-related since decarbonizing the grid relies on scenarios that assume increased penetration of variable generation technologies that produce electricity dependent on climatic conditions including wind, photovoltaic, hydro, combined heat and power and tidal technologies. As penetration rates of variable generation increase to 15–20 per cent, depending on the electricity system, it will become increasingly difficult to ensure the reliability of electricity systems based on conventional grid architectures and limited flexibility (IEA, 2011). Smart grids can potentially support greater deployment of variable generation technologies by providing operators with real-time system information that enables them to manage generation, demand and power quality, thereby increasing system flexibility and maintaining stability (Charles, 2009).

Although electricity consumption represents 17 per cent of final energy use today, it causes 40 per cent of global $CO_2$ emissions because nearly 70 per cent of electricity is produced from fossil fuels (IEA, 2010). If decarbonization measures are implemented, the contribution of electricity generation to global $CO_2$ emissions could fall to 21 per cent by 2050. And when taking direct and indirect emissions reductions into account, smart grids could potentially reduce net annual $CO_2$ emissions by 0.7 to 2.1 Gt by 2050 (IEA, 2010).

New research is considering how the smart grid concept can be extended to broader energy and societal contexts termed the 'smart community' or 'smart city'. This involves integrating several energy supply and use systems within a region to optimize operations, while maximizing the integration of RE resources (from large-scale wind to micro-scale photovoltaic) and coupled with residential energy management systems. This would also include existing infrastructure systems, that is, electricity, water, transportation, gas, waste and heat, and future systems, that is, hydrogen and electric vehicle charging (IEA, 2011).

## CONCLUSIONS

This chapter has assessed the key opportunities and barriers for shifting towards a more sustainable low carbon transport system. At present, the largest energy users are LDVs. While there are large challenges, there are also important opportunities that will depend on taking an integrated approach. For instance, large-scale vehicle electrification to meet carbon emission targets is dependent upon decarbonized electricity while ensuring sufficient capacity for increasing demand (CCC, 2008; Kannan, 2009; Skea et al., 2011). Another important issue is that climate policy will likely converge with the energy security agenda. Risk and uncertainty in global energy markets will influence energy policy because the depletion of domestic fossil fuel resources will increase reliance on imports from increasingly concentrated and politically unstable suppliers (DTI, 2007; Nakhle, 2007).

Due to nearly complete reliance on liquid fossil fuels, transport is the most vulnerable to energy security risk. Yet, global trade in fossil fuels is expected to double by 2030, potentially increasing the risk of price volatility and supply disruptions due to a decline in spare crude oil production capacity in supplier countries, lower levels of inventories and a rapid decline in surplus capacity in the global refining industry. There are also many issues surrounding civil unrest and political instability in exporter regions that may affect global oil markets. Consequently, security of supply will be affected by a tightly balanced market more sensitive to actual or perceived threats to supply disruptions, where swings in demand are increasingly met by price changes rather than delivery from storage (Costantini et al., 2007; DTI, 2007). Measures including improving vehicle efficiency, displacing liquid transport fuels with low carbon grid electricity, increasing use of renewables such as biofuels and hydrogen and reducing oil price inelasticity in the transport sector can potentially reduce the market risks associated with price volatility and supply disruption (Costantini et al., 2007; WRI, 2007; Gallagher and Collantes, 2008) while shifting towards a lower carbon transport system.

## NOTES

1. G8 countries: Canada, France, Germany, Italy, Japan, Russia, UK and USA. The other five countries are leading emerging economies and include: Brazil, China, India, Mexico and South Africa.
2. The category of LDV usually includes cars and light trucks, including minivans and sport utility vehicles (SUVs).
3. Lignocellulosic biomass can be grouped into four main categories: agricultural residues

(including corn stover and sugarcane bagasse), dedicated energy crops, wood residues (including sawmill and paper mill discards) and municipal paper waste.

# REFERENCES


Boots, M., D. Thielens and F. Verheij (2010), *International Example Developments in Smart Grids – Possibilities for Application in the Netherlands*, Arnhem: KEMA Nederland B.V.

Campbell, J.E., D.B Lobell and C.B. Field (2009), 'Greater transportation energy and GHG offsets from bioelectricity than ethanol', *Science*, **324**(5930), 1055–7.

CCC (Climate Change Committee) (2008), *Building a Low-carbon Economy – The UK's Contribution to Tackling Climate Change*, London: The Stationery Office.

Charles, D. (2009), 'Renewables test IQ of the grid', *Science*, **324**(5924), 172–5.

Costantini, V., F. Gracceva, A. Markandya and G. Vicinia (2007), 'Security of energy supply: comparing scenarios from a European perspective', *Energy Policy*, **35**(1), 210–26.

DOE (Department of Energy) (2009), *Smart Grid System Report*, Washington, DC: US Department of Energy.

DTI (Department of Trade and Industry) (2007), *Meeting the Energy Challenge: A White Paper on Energy*, London: UK Department of Trade and Industry.

Gallagher, K.S. and G. Collantes (2008), *Analysis of Policies to Reduce Oil Consumption and Greenhouse-Gas Emissions from the U.S. Transportation Sector*. Cambridge, MA: John F. Kennedy School of Government, Harvard University.

IEA (International Energy Agency) (2008), *Energy Technology Perspectives*, Paris: IEA.

IEA (International Energy Agency) (2009), *Transport, Energy and $CO_2$: Moving Towards Sustainability*, Paris: IEA.

IEA (International Energy Agency) (2010), *World Energy Outlook*, Paris: IEA.

IEA (International Energy Agency) (2011), *Technology Road Maps – Smart Grids*, Paris: IEA.

IPCC (Intergovernmental Panel on Climate Change) (2007), *Synthesis Report*, Working Group 3: Mitigation of Climate Change, IPCC.

IPCC (Intergovernmental Panel on Climate Change) (2011), *Special Report Renewable Energy Sources (SRREN)*, Working Group 3: Mitigation of Climate Change, IPCC.

Jacobsson, M.Z., W.G. Colella, D.M. Golden et al. (2005), 'Cleaning the air and improving health with hydrogen fuel-cell vehicles', *Science*, **308**(5730), 1901–5.

Kannan, R. (2009), 'Uncertainties in key low carbon power generation technologies – implication for UK decarbonisation targets', *Applied Energy*, **86**(10), 1873–86.

King, J. (2007), *The King Review of Low Carbon Cars*, London: HM Treasury.

Lopez, J.M., A. Gomez, F. Aparicio and J. Sanchez (2009), 'Comparison of GHG emissions from diesel, biodiesel and natural gas refuse trucks in the City of Madrid', *Applied Energy*, **86**(5), 610–15.

Nakhle, C. (2007), 'Do high oil prices justify an increase in taxation in a mature oil province? The case of the UK continental shelf', *Energy Policy*, **35**(8), 4305–18.
Ogden, J.M., R.H. Williams and E.D. Larson (2004), 'Societal lifecycle costs of cars with alternative fuels/engines', *Energy Policy*, **32**(1), 7–27.
O'Neill, B.C., M. Dalton, R. Fuchs, L. Jiang, S. Pachauri and K. Zigova (2010), 'Global demographic trends and future carbon emissions', *Proceedings of the National Academy of Sciences of the USA*, **107**(41), 17521–6.
Owen, N.A., O.I. Inderwildi and D.A. King (2010), 'The status of conventional world oil reserves – hype or cause for concern?', *Energy Policy*, **38**(2010), 4743–9.
Samaras, C. and K. Meisterling (2008), 'Lifecycle assessment of greenhouse gas emissions from plug-in hybrid vehicles: implications for policy', *Environmental Science and Technology*, **42**(9), 3170–6.
Service, R.F. (2004), 'The hydrogen backlash', *Science*, **305**(5686), 958–61.
Skea, J., P. Ekins and M. Winskel (eds) (2011), *Energy 2050: Making the Transition to a Secure Low Carbon Energy System*, London: Earthscan.
UKERC (UK Energy Research Centre) (2009), *Global Oil Depletion – An Assessment of the Evidence for a Near-term Peak in Global Oil Production*, London: UK Energy Research Centre.
UN (United Nations) (2005), *World Urbanization Prospects: The 2005 Revision Population Database*, New York: UN Department of Economics and Social Affairs.
UN (United Nations) (2007), *World Population Ageing 2007*, New York: UN Department of Economics and Social Affairs.
UNFPA (United Nations Population Fund) (2007), *State of World Population 2007: Unleashing the Potential of Urban Growth*, New York: UNFPA.
UNPD (United Nations Population Division) (2009), *World Population Prospects: The 2008 Revision*, New York: UNPD.
UNPD (United Nations Population Division) (2010), *World Urbanization Prospects: The 2009 Revision,* New York: UNPD.
WRI (World Resources Institute) (2007), *A Snapshot of Selected U.S. Energy Options Today: Climate Change and Energy Security Impacts and Tradeoffs in 2025*, Washington, DC: WRI.

# 9. Finance and investment in transport

Nihan Akyelken

## INTRODUCTION

Historically, transport infrastructure investment is considered to be a precondition for an efficient economy. Given the large costs associated with infrastructure projects, it has long been recognized that transport is costly; therefore, public finances alone are not adequate. This, in turn, has determined the financing options available for transport. The world is in an era of transition however. Climate change, depleted energy resources, globalization and population forces act as significant pressures on the global economy. To tackle these challenges, new societal visions are being formed, including de-growth, green growth and neo-growth (Banister, 2012b). Efforts towards a reconceptualization of growth have also abounded: the quest for a better indicator of economic progress and wellbeing has become a prevalent trend (Stiglitz Commission, Stiglitz et al., 2009).

During such complex and profound transformation the role of infrastructure development in growth diminishes, entailing that sustainability elements be incorporated into transport. This has significant implications for types of investments and financing tools. The objective of achieving low carbon mobility has wide-ranging implications for finance and investments in transport. Once transport is expected to play a role in the low carbon growth agenda, the goal of transport investment is no longer limited to economic growth and increased productivity. Accordingly, current transport investment and finance decisions should meet the increasing demand for mobility while ensuring that the mobility patterns are sustainable. The good news is this is not impossible as long as a holistic approach to investments is taken. Indeed, it may even generate more financing opportunities, ensuring both economic and environmental sustainability.

Within this broad setting this chapter reviews the current and future challenges involving finance and investment in transport in the context of low carbon mobility. Two problems emerge from previous research.

Conceptually, the literature on economic growth and the literature on the sustainability impacts of transport investments have been disconnected. At an operational level, the long-standing focus on (road) transport infrastructure investments with the aim of economic growth has influenced the way financing options are designed. Furthermore, risks associated with private and public financing options, and how they exert pressure on or point to new opportunities for inclusion of sustainability in transport investments are worth noting. In light of this, the second section of the chapter presents the implications of these problems and opportunities for low carbon mobility, focusing on recent trends revealed in the existing research.

Next, a brief overview of the literature on the economic growth impacts of transport investments is presented in order to highlight the conceptual problem inherent in the detachment of the economic and environmental sustainability aspects of transport investments. Then, the key issues in financing investments are highlighted with the aim of emphasizing the narrow approach to financing options. In the second section of the chapter three key trends are highlighted: the diversification of investments across different types of transport modes; internalization of external costs; and the role of climate change finance in transport. Based on this and drawing on the existing research, suggestions on how to overcome the issues of finance and investment in low carbon mobility are presented.

## KEY RESEARCH THEMES IN TRANSPORT FINANCE AND INVESTMENT

### Economic Growth and Transport Investment

The direct benefits of transport investments include travel time and cost savings as well as short-term job creation effects mainly in the construction sector. Indirectly, it is deemed to have a facilitator role, as accessibility increases productivity, which results in increased trade activity and hence provides cohesion between the regions through reduced transport costs (Goetz, 2011). In fact, such wider benefits of transport infrastructure development have recently gained a place in the policy agenda (for example, the UK Department for Transport appraisal scheme's inclusion of wider benefits) following the evidence provided by several studies including Graham (2007) and Gibbons and Overman (2009). These studies mainly pointed to increased productive capacity through external economies of scale, including the market accessibility of producers and the stimulation of private investment in the context of agglomeration economies. More broadly, such indirect impacts of transport investment on economic development have been tackled in different research strands, including development theory, economic geography, transport geography and transport economics. Among the earlier studies development theorists (for example, Nurkse, Rosenstein-Rodan, Hirschman, cited in Krugman, 1998) emphasized that investment in transport is the key starting point for the industrialization process, as it exhibits increasing returns to scale.

Interestingly, although these wider economic benefits of transport infrastructure investment were identified early in the development and economic geography literature, subsequent empirical research on transport investment impacts has mainly focused on the direct benefits of transport, mostly through cost-benefit analyses (Lakshmanan, 2011). The transport/development economics literature took the lead in empirically testing the relationship between the two, an approach that became popular following a seminal paper by Aschauer (1989). In this paper he examined public infrastructure productivity using a production–function approach. Aschauer's (1989) findings on the positive effect of public infrastructure on aggregate production inspired several scholars, resulting in empirical investigations in different geographical contexts (see Goetz, 2011 for an overview). The methodologies of the subsequent macroeconomic studies are diverse (see Torrisi, 2009 for a broad overview). The criticism against these studies has also been vast. One of the fundamental criticisms was the problem of causality. The causality argument implies that the existence of a statistical relationship between productivity and investment in public infrastructure does not explain in which direction the causality runs (Gramlich, 1994; Garcia-Milà et al., 1996). These issues have led to more sophisticated econometric models: introducing a time lag between variables (that is, using a vector auto regression (VAR) model). Demetriades and Mamuneas (2000) and Esfahani and Ramírez (2003) addressed the causality issue, and showed that transport investments have a positive impact on economic growth.

However, the variation between the results of these macroeconomic models was striking (Goetz, 2011; Lakshmanan, 2011). Bhatta and Drennan (2003) and Goetz (2011) conducted reviews of the empirical literature on the economic benefits of public investment in transport during the periods 1989–99 and 1999–2009, respectively. The overarching implication of these reviews is that, because the number of empirical studies that found a positive impact outnumbered those that found either insignificant or negative impacts, it cannot be concluded that transport infrastructure investment is beneficial for economic development in all regions at all phases of development. In fact, because the variation among the results is significantly wide, the reviews conclude that the economic development implications of transport infrastructure investment should be considered in geographical and historical contexts.

In the last two decades assertions that transport infrastructure is either a necessary, but not sufficient, condition for development or that it is neither necessary nor sufficient have appeared in the literature (Banister and Berechman, 2000). This is mainly rooted in the idea that there is a need for further factors that complement infrastructure. A holistic understanding of transport investments that incorporates social and environmental sustainability is thus required. However, the economic growth impacts of transport investments and sustainability in the transport literature have remained detached from each other (Goetz, 2011). While the former research strand aims at elucidating the direct links between investment in transport infrastructure and productivity as explained above, the latter points out the external costs of transport, which are not internalized by transport users (Hensher and Button, 2003). Despite the very large extent of both literatures, there are only a few studies (for example, Feitelson, 2002; Stead and Banister, 2006) that have attempted to fill this gap by pointing to the need to consider the negative externalities associated with transport investment decisions, as well as the intentions of economic growth.

## Public Versus Private Finance and Risk-sharing

Traditionally, governments have been the main sources of funding for transport infrastructure. As stated earlier, it has long been acknowledged that public budgets are not adequate to cover the growing number of transport infrastructure projects; private sources are therefore considered to be essential (Small, 2010). Sweden provides an example of private financing in transport: 67 per cent of the total length of roads in Sweden is private (ITF, 2008). In general, the most common use of private funding for transport infrastructure is in the form of a delegation of responsibilities to a private partner over a long period of time, namely, public private partnerships (PPPs). The theoretical rationale behind PPPs is that competition induced as a result of involving private parties renders the projects more effective. PPP contracts can take many forms: the government may design the project and delegates construction responsibilities to private parties or lets the private parties design, build, construct, operate and maintain the project through the provision of concessions and so on (see Rodrigue, 2012 for a brief overview of PPP types). The use of PPPs in transport is popular at a global level: in the period 1985–2008 over 1100 projects were funded by PPPs worldwide (AECOM, 2005, cited in Siemiatycki, 2010).

The existing research has identified several reasons for the increased use of private finances in transport, including PPPs in addition to inadequate public finances. These include the positive experiences in the use of private sector finance in other sectors (including water and energy), the effective pricing strategies employed in the private sector, the variety of public sector funding sources (for example, pensions funds), increased efficiency, the profit motive for the private sector and the ease of risk-bearing (McQuaid and Sherrer, 2010; Small, 2010). In the context of urban transport, cross-subsidization is also considered to be a pressing issue: revenues from user fares are usually inadequate to maintain transport services so that government subsidies are required. This is evident in American transit systems, where a major share of the funding in public bus transport systems in major states comes from subsidies (Rodrigue, 2012). At the macro level, it was empirically shown by Hammami et al. (1999) that PPPs tend to be more common in countries whose governments suffer from heavy debt burdens and where aggregate demand is sizable. However, this claim should be re-evaluated considering the current fiscal crises in Europe. Finally, market size and market demand, as captured by population and real gross domestic product (GDP) per capita, as well as the legal system channel, were found to be particularly important in promoting the use of transport PPPs (Hammami et al., 1999).

However, most of the arguments for and drivers behind private sector involvement in transport investments have been problematized both theoretically and empirically. Issues relating to pricing and franchising in a competitive economy have significant negative implications for society. First of all, as with all other free market economic structures, the risk of monopoly power use (that is, setting high mark-ups above market prices) leads to distribution problems (Small, 2010). Also, governments' tendering of the contracts to the private sector involves significant transactions costs, which can be divided into *ex ante* and *ex post* costs; the former refers to information problems, while the latter includes problems regarding long-term projections (that is, the pricing projections over the long term are unreliable) (Small, 2010). For instance, through a case study on the UK, Shaoul et al. (2006) showed that the cost of transfers between the government and the private sector could be higher than the cost of public investment in the infrastructure.

Ownership structures of transport projects also entail issues of risk-sharing. For reasons stated above, infrastructure projects are risky, and these risks arise in all stages of the transport projects. Among the risks listed by Checherita and Gifford (2007) the existing research has provided the most extensive evidence for technical risks involved in the design of the projects and the potential cost overruns. In this research strand the works of Flyvbjerg et al. (2003, 2004) are of key importance. They compiled a dataset of 258 infrastructure projects including rail (58), fixed links (33) and roads (167) completed between 1927 and 1998. Almost 95 per cent of these projects were carried out in the developed world, mainly in Europe and North America. Using this data Flyvbjerg et al. (2003, 2004) investigated the differences between the actual costs and the forecasted costs of these projects together with the causes of the differences. The results are striking: they found that the actual costs are on average 28 per cent higher than the forecasted costs. This goes up to 45 per cent for rail projects and 34 per cent for fixed links, while for road transport projects the average cost escalation is 20 per cent.

Moreover, the results also showed that cost escalation has not decreased over the past 70 years, indicating that no social learning has taken place. Flyvbjerg et al. (2004) also looked at the causes of such high levels of cost escalation and found that the length, type and ownership of the projects are important determinants of cost escalations, and that cost escalation appears to be more pronounced in developing nations than in North America and Europe, although the analysis was carried out only for rail due to data availability. These results, however, contradict the results of Gamez and Tauran (2010), who found that the costs for World Bank-funded transport projects in developing countries are usually overestimated: most projects finish under budget.

On the whole, key research themes in transport investment and finance included economic growth and sustainability impacts of transport separately. Moreover, the research that looked at the risks involved with different types of transport projects is worth noting. Against this research setting, the key trends and issues involved in investing in and financing low carbon mobility are highlighted in the following section.

## LOW CARBON MOBILITY: THE KEY ISSUES

### Investment in Alternative Transport Modes

Past transport investment trends with the aim of increasing economic growth – particularly ones focusing on road transport infrastructure investment in developing countries – have led to the current unsustainable transport patterns (Sakamoto et al., 2010; see also Chapter 1). The induced demand effects of road building is also an important factor: the induced travel as a result of increased road infrastructure leads to even more investment in road transport and so eventually to higher levels of negative environmental effects (SACTRA, 1999). Furthermore, the global finance that is raised for developing countries has also mainly focused on road transport, including those transport investments provided by overseas development agencies. The same holds for export credits that inject foreign direct investment into developing countries with the aim of economic growth (Sakamoto et al., 2010). The investment trends in the European Economic Area (EEA-32) illustrate this dominance by road transport investments. Based on European Environment Agency (EEA) (2011) data, investment in transport infrastructure for the EEA-32 Member States rose steadily between 1998 and 2008 (Figure 9.1). As of 2008, it accounted for 1.2 per cent of GDP. The EEA divides this period into two: the first five years' growth in investment was driven by an increase in rail infrastructure spending (from 0.25 per cent to 0.38 per cent of GDP) while the latter period saw a rise in spending on road infrastructure (from 0.57 per cent to 0.70 per cent of GDP). Despite the increase in spending on rail, spending on roads still constitutes the major share of transport infrastructure investment in the EEA-32.[1]

However, recent investment trends indicate that this trend is unlikely to hold both in passenger and freight transport. When coupled with the impacts of the financial crisis this potentially has a positive signal for shifts to alternative modes. First, the European Union (EU) has ambitious targets for the increase of rail freight and intermodal transport, as made evident in the 2011 EU Transport White Paper's emphasis on intermodality and modal shift; this has the potential to divert finances from road to alternative modes of transport. In fact, the recent TEN-T guidelines published in October 2011 suggest the implementation of a corridor development approach in the core network, where each corridor will include three modes, three Member States and two cross-border sections and provide corridor platforms as a governance structure (EC, 2011b). This will mean better management of cross-border projects for a multi-modal network. In addition to the EU-wide targets, the recent investments in high-speed rail in both developed and emerging economies also confirm this trend. For instance, most recently Poland received around 224 million euros from the European Investment Bank, which will buy 20 trains with a top speed of about 250 km per hour (EC, 2011c). These modal shift developments also reach beyond the EU: in the US President Obama plans to provide 80 per cent of American residents with access to high-speed rail within the next 25 years. To that end, the Federal Railroad Administration has already solicited applications for more than eight billion euros in grant funding (FRA, 2009). Recently, the California Assembly approved funding for 209 km of the high-speed rail line that is expected to link Los Angeles to San Francisco (BBC News, 2012).

Second, the impacts of the financial crisis may have budgetary implications regarding investments in different modes. Bowen and Stern (2010)

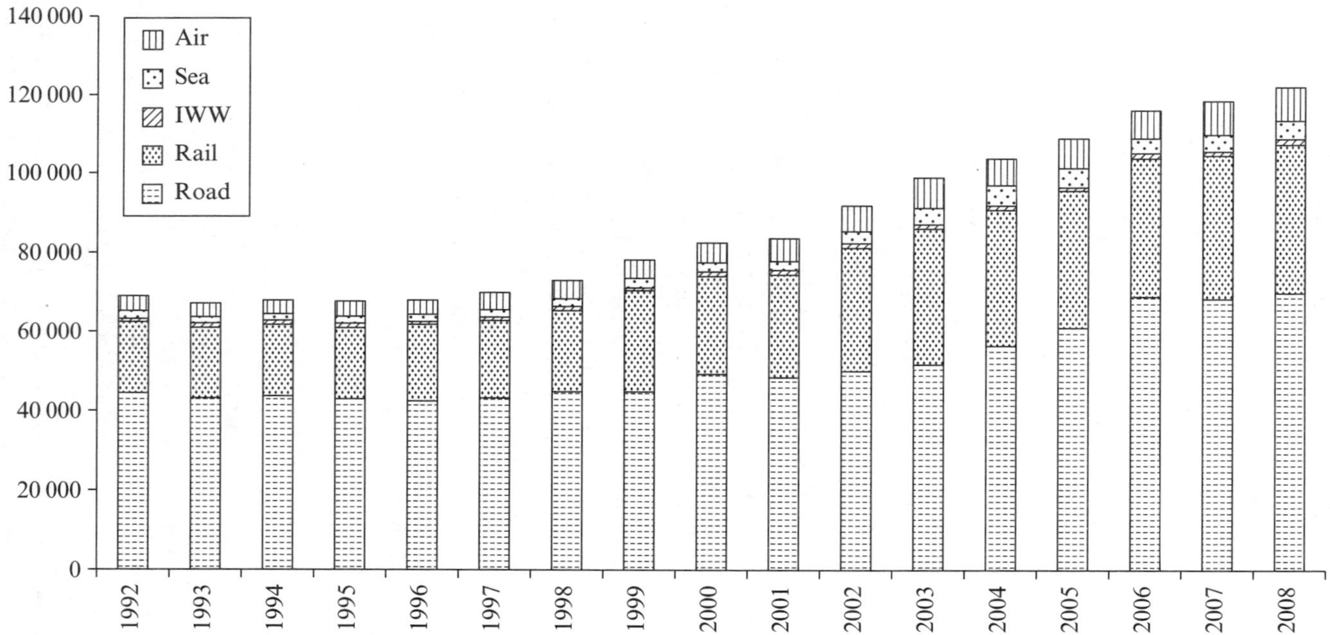


*Note:* IWW – inland waterways.

*Source:* EEA (2011).

*Figure 9.1* *Investments in infrastructure (million euros) by mode: EEA-32*

argue that it could well be that although the budget for the environmental measures may be reduced, the proportion devoted to rail infrastructure may increase, which would have a positive impact on the environment through a possible modal shift from road to rail; and in fact this is already happening as stated above. All these recent pieces of legislation, policy goals and increased investment in alternative modes of transport have the potential to make it easier to incorporate low carbon mobility elements into transport finance and investment.

However, two issues are worth noting. First, road transport projects are still popular even in advanced countries including Western Europe. Although it does include railway projects (for example, in the UK the electrification of the Transpennine Express link and the railway link between Oxford and Bedford), the National UK Infrastructure Plan 2011 (HM Treasury, 2011), for instance, secured £1 billion to tackle areas of congestion and improve the national road network. Second, the environmental impact of using alternative modes of transport such as rail would depend on the fuel mix in the rail network: if the rail network is electrified, and electricity is generated from renewable sources, the changes in environmental emissions would be substantial. If not, however, the outcome is uncertain, particularly when the emissions from building the rail network are included in the calculations (Bowen and Stern, 2010). For example, Chester and Horvath (2010) found that although high-speed rail in California has the potential to reduce greenhouse gas emissions and energy consumption, it may actually increase sulphur dioxide ($SO_2$) emissions, which would lead to environmental acidification and eventually health issues. Again, they stress that this depends on the electricity mix used. In that vein, in terms of increasing the use of renewable energy, funding and incentives provided by the national governments is also significant because market forces alone will not achieve this given the existing infrastructure and the existing vehicle types (see Chapter 8).

Moreover, wider soft policy measures aimed at sustainable travel and carbon reduction at city and national levels both have mode-shift and sustainability impacts, and require much less funding. For instance, policy measures to encourage walking and cycling (for example, funding walking and cycling schemes), urban planning (for example, investment in walking and cycling facilities) and the use of information and communication technologies in car-sharing and public transport information systems alongside increasing the frequency of public transport are relatively inexpensive to governments and local councils (Hickman et al., 2010). According to the findings of the VIBAT London study (http//www.vibat.org), which considers the role of the transport sector in reducing carbon dioxide ($CO_2$) emissions in London, ecological driving and slower speeds have the potential of 15–20 per cent $CO_2$ reduction in London. All these measures can be easily funded and encouraged by national governments as they do not require such high levels of funding as large infrastructure projects.

### Internalization of Externalities

In economic theory externalities are defined as the costs and benefits to third parties that are not involved in the transaction, that is, they are not priced. Accordingly, the (environmental) externalities can be addressed by taxing or subsidizing the externality depending on whether it is a positive or a negative externality. However, most of the transport-demand measures, pricing in particular, are intended to raise revenue rather than address negative externalities (Button and Vega, 2008). Santos et al. (2000) have shown that road charging has environmental benefits if it is applied only on certain cordons at certain times to benefit from the energy-reduction effects of decreasing congestion and perhaps reducing traffic volume. Moreover, some (for example, Van der Bergh et al., 2007) argue that the inclusion of externalities in transport infrastructure projects/pricing might hinder innovations by raising the cost to producers, while some (for example, Evenhuis and Vickerman, 2010) point out the uncertainty problems for the revenues of the private party from the internalization of externalities in transport projects. On the government side, there are deemed to be real commitment problems as government promises over lengthy periods of time lack credibility (Thatcher, 2008). Evenhuis and Vickerman (2010) also point out that internalization is even more challenging for roads, railways and inland waterways projects, as, unlike ports and airports, they are significantly affected by network effects.

Despite these problems with the incorporation of externalities in transport projects, Evenhuis and Vickerman (2010) suggest that performance-based pricing by the government should solve some problems: put simply, the users pay the government and the government pays the private party. Another theoretical illustration of road pricing on environmental externalities (for example, less fuel consumption and mode-shift behaviours) was carried out by De Jong et al. (2010). They have shown that a distance-based charge of 0.15 euros per km can decrease the share of road transport and induces mode-shift impacts as well as less fuel consumption (Figure 9.2).

Like its theoretical exploration, the empirical illustration of the internalization of external costs is diverse. For instance, purchase taxation measures have recently been reformed to promote cleaner and low carbon vehicle technologies (Potter, 2008). Indirect substitutions by governments are also expected to have environmental benefits, with the vehicle

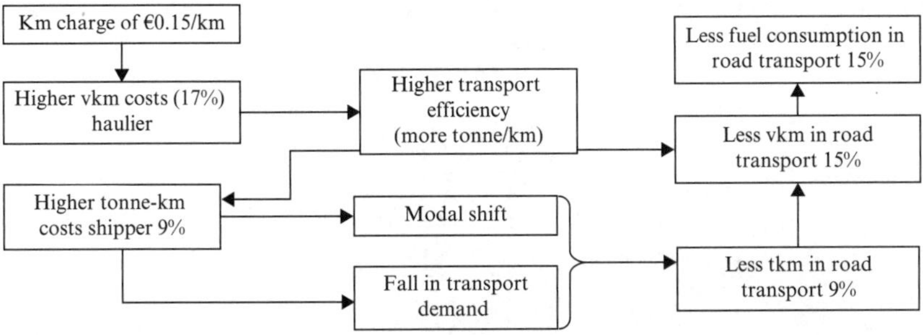


*Source:* De Jong et al. (2010).

*Figure 9.2 Impacts of distance-based charging*

scrappage programmes in many EU countries promoting the replacement of old vehicles with fuel-efficient vehicles (UNEP, 2011). On the whole, the central environmental motives behind pricing/taxing/subsidizing can be summarized as mode shifts, congestion reduction and sustainable energy consumption. Mode-shift effects are considered to be particularly important as they have been shown to be less costly than fuel technologies in terms of carbon reduction from transport, that is, the cost of a fuel-based solution is at least ten times more than the cost of shifting modes (Wright and Fulton, 2005).

## Climate Change Finance and Transport

Finally, that climate-oriented financial mechanisms have not adequately covered transport is of vital importance for exploring ways of financing low carbon mobility. As of April 2012, among 8128 Clean Development Mechanism[2] (CDM) projects only 47 transport projects were funded (Figure 9.3). These transport projects include bus rapid transit; mode-shift, bio-diesel for transport projects; motorbikes; rail: generative breaking; and metro-efficient operation. The reduction impacts of these projects are reported whenever they are available. For instance, installation of low greenhouse gas-emitting rolling stock cars in the metro system in India in 2007, which is one of the earliest transport projects funded by CDM, reduced approximately 42000 tonnes of $CO_2$ per year (The Clean Air Institute, 2008). In Mexico City a 10 per cent reduction in certain environmental emissions, the result of the introduction of bus-based rapid transit systems, bikeways, climate-friendly technologies and traffic measures, was estimated to yield an average annual benefit of $759 million (World Bank, 2006). Moreover, the welfare gain from improved air quality is expected

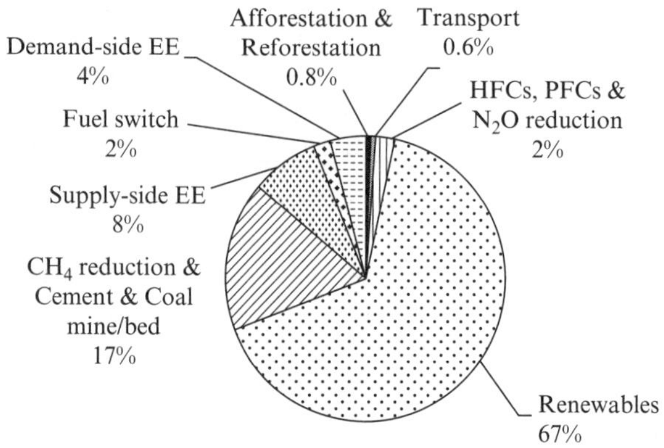


*Note:* EE – energy efficiency; $CH_4$ – methane; $N_2O$ – nitrous oxide; HFCs – hydrofluorocarbons; PFCs – perfluorocarbons.

*Source:* UNEP (2012).

*Figure 9.3 Sectoral breakdown of CDM*

to result in an estimated annual benefit of $2 billion (World Bank, 2006). However, it should be noted that quantifying the environmental and social impacts of policies through an economic paradigm can be based on strong assumptions that may yield misleading conclusions. Re-examining the appraisal methods to avoid such issues for transport infrastructure could be carried out in line with green growth strategies, that is, the 'avoid–shift–improve' strategy (see Chapters 1 and 13).

Moreover, it is not always possible to identify all emissions from transport and to specify the baseline scenario and project boundaries, which make the transport sector incompatible with CDM's project-based approach (Banister, 2012a). Compared to other sectors, transport projects also require high levels of lump-sum payments. Overall, it is said to be almost impossible to account for the complexities of urban transport in the existing CDM project guidelines, to quantify the wider benefits of transport in particular (Banister, 2012). At an operational level, the CDM barrier analysis used to identify the financial barriers for the project implementation is not found to be convincing for transport as the carbon credits tend to represent less than 2 per cent of the infrastructure investment and the analysis cannot demonstrate that the project would not have been carried out without the CDM funding (Sanchez, 2008).

The unsuitability of the transport sector within the CDM framework has attracted significant attention: several suggestions have been offered to overcome barriers in the use of the CDM for transport. The fundamental message of the existing suggestions is to provide a bottom-up and a sector-based approach implying the use of Nationally Appropriate Mitigation Actions (NAMAs) for city-wide application – a view also shared by the United Nations (Banister, 2012a). According to this view, NAMAs are useful to provide policy guidelines and programmes for national governments to implement in the transport sector. What makes NAMAs more appropriate for the transport sector is that they are not subject to financing barriers that appear in the use of emissions certificates in transport (Helme, 2010). Other tools like Environmental Fiscal Reforms (EFRs) that shifts taxation from labour to environmental use are also found favourable for the transport sector as it provides sector-specific guidelines that are connected with carbon markets (Banister, 2012b). In addition to NAMAs and EFRs, the special emphasis of the World Bank's latest Global Environment Facility (GEF-5 – a flexible funding mechanism under the United Nations Framework Convention on Climate Change (UNFCC) that covers the period between 2010 and 2015) on investment in low carbon transport is worth noting. The focus on land transport in these funding mechanisms has now been expanded to cover transport and land planning and the funds have concomitantly been increased to $250 million, which is expected to yield $1.3 billion (STAP, 2010). On the whole, 6.4 per cent of total investment is devoted to transport projects (Banister, 2012b). An example of the use of GEF in transport is the building of the 50 km metrobus system in Mexico City, which can reduce $CO_2$ emissions from Mexico City traffic by an estimated 60000 to 80000 tonnes a year, and has removed 800 polluting minibuses from the road and encouraged walking and cycling (GEF, 2012). These figures require further attention given different methodologies employed.

Most recently, the CDM Policy Dialogue (High-level Panel on the CDM Policy Dialogue, 2012) has taken into account these issues in its recommendations across 12 areas for enhancing CDM. In the report the methodological concerns about transport and CDM are confirmed by the stakeholders' views warning that the current coverage of methodologies may hinder sustainable development potential of the transport sector. This is reflected in the report's recommendation to encourage greater access to CDM for underrepresented regions: it is recommended that the CDM Executive Board should accelerate the development of standardized parameters and simplified procedures for providing public services like mass transport.

Although the carbon market is found not suitable for transport, it is also worth noting the implications of the exclusion of transport in carbon-pricing initiatives. The absence of transport in the current carbon-trading schemes (for instance, in the EU Emissions Trading Scheme (ETS)) has rendered emissions from transport much more visible compared to other sectors. According to the latest EU figures, emissions covered by the EU ETS between 2008 and 2009 decreased by 11 per cent, following a cut of 6 per cent the year before, while emissions from transport have continued to grow. However, recent developments have integrated parts of transport in the EU ETS. Aviation is included from 1 January 2012 onwards. Also, in the shipping sector there have been proposals to adopt an ETS system. The European Commission (EC) has also looked at the possible inclusion of road transport but has focused on inclusion of car manufacturers and of individual motorists. The effectiveness of the ETS on transport in terms of raising finances for transport or reducing its environmental impact is still unknown (Kritzinger et al., 2011).

## CONCLUSION

Previous research had indicated that we may be facing a dilemma: is there a trade-off between the goals of investment and finance in transport and should these goals be to increase economic growth or to ensure environmental sustainability? Fortunately, recent trends and studies show that this is not necessarily the case. In the past, financing and investment trends have been mainly focused on road infrastructure investment, which has had a significant impact on current unsustainable trends in transport. Furthermore, the fact that the transport sector did not receive a large share from the climate change finance and other carbon-pricing options has made the situation more difficult. The recent emphasis on the use of internalization of externalities in private and public financial flows, together with the efforts on increasing the share of transport in global climate change funds, are green lights for raising finance for sustainable transport. However, the empirical evidence on past transport projects and the lack of sustainability elements in the current appraisal methods require significant attention. Unfortunately, it is very likely that the existing assessment methods may yield misleading conclusions for future developments.

The key message is clear: profound conceptual transformation is required. This must include adopting the options to reduce energy/$CO_2$ intensity of growth or, more radically, thinking of a different way of measuring progress in society, a way that incorporates sustainability measures into the understanding of growth. Specifically, this chapter has shown that a holistic solution is required, especially in developing countries, which includes solutions both within the transport sector and in climate change finance in general. The efforts made towards the incorporation of externalities in public/private funding and PPPs and other pricing/taxation measures should help to improve finance and investment related to low carbon mobility. In short, with the increasing role of transport in low carbon growth strategies, sustainability elements can be well integrated with transport investment and financing decisions. By also encompassing an increase in the share of sustainable transport funds, a holistic view of an economically and environmentally sustainable transport system becomes possible.

## NOTES

1. It is worth noting that not all spending has targeted building new infrastructure. In fact, a new move has emerged: building new transport infrastructure has started to be replaced by rebuilding and maintaining the existing ones. The recent political discussions in the US calling for a 'fix-it-first' approach in American highways as part of the smart growth agenda illustrate this. Kahn and Levinson (2011) discussed that the US is currently underinvesting in maintaining existing transport infrastructure and overinvesting in other, less productive new infrastructure. Looking at the recent investments in urban areas, the move towards improving the existing infrastructure becomes apparent: $40 686 000 was recently invested in repairing 86 miles of roads and six bridges in Long Island in April 2012 (New York State Press Office, 2012).
2. As defined in Article 12 of the Kyoto Protocol, 'CDM allows a country with an emission-reduction or emission-limitation commitment under the Kyoto Protocol (Annex B Party) to implement an emission-reduction project in developing countries' (United Nations Framework Convention on Climate Change website, http://unfccc.int/kyoto_protocol/mechanisms/clean_development_mechanism/items/2718.php) (accessed 10 July 2012).

## REFERENCES


Aschauer, D.A. (1989), 'Is public expenditure productive?', *Journal of Monetary Economics*, **23**, 177–200.

Banister, D. (2012a), 'Urban mobility and the environment', in *Global Report on Human Settlements 2013*, UN-Habitat, forthcoming.

Banister, D. (2012b), 'Introduction to the future of Europe in a global world proposal', Unpublished, Transport Studies Unit, University of Oxford, Oxford.

Banister, D. and J. Berechman (2000), *Transport Investment and Economic Development*, London: University College London.

BBC News (2012), *Californian High-speed Rail Wins Approval*, available at http://www.bbc.co.uk/news/world-us-canada-18748717 (accessed July 2012).

Bhatta, S.D. and M.P. Drennan (2003), 'The economic benefits of public investment in transportation – a review of recent literature', *Journal of Planning and Education and Research*, **22**(3), 288–96.

Bowen, A. and N. Stern (2010), 'Environmental policy and the economic downturn', *Oxford Review of Economic Policy*, **26**(2), 137–63.

Button, K. and H. Vega (2008), 'Road user charging', in S. Ison and T. Rye

(eds), *The Implementation and Effectiveness of Transport Demand Management Measures*, Hampshire: Ashgate Publishing, pp. 29–48.
Checherita, C. and J. Gifford (2007), *Risk Sharing in Public–Private Partnerships: General Considerations and an Evaluation of the U.S. Practice in Road Transportation*, Washington, DC: Transportation Research Forum 2007.
Chester, M.V. and A. Horvath (2010), 'Assessment of high-speed rail: the case of California', *Environmental Research Letters*, **5**(1), 1–8.
De Jong, G., A. Schroten, H. Van Essen, M. Otten and P. Bucci (2010), *Price Sensitivity of European Road Freight Transport – Towards a Better Understanding of Existing Results*, Report for Transport and Environment, The Hague.
Demetriades, P.O. and T.P. Mamuneas (2000), 'Intertemporal output and employment effects of public infrastructure capital: evidence from 12 OECD economies', *Economic Journal*, **110**, 687–712.
EC (European Commission) (2011a), *Roadmap to a Single European Transport Area – Towards a Competitive and Resource Efficient Transport System*, COM (2011) 0144 (Final), 28 March, Brussels: European Commission.
EC (European Commission) (2011b), *A Growth Package for Integrated European Infrastructures*, COM(2011) 676/2, Brussels: European Commission.
EC (European Commission) (2011c), *Poland: New High-speed Trains for Railway – EIB lends EUR 224 m to PKP Intercity, BEI/11/113*, available at http://europa.eu/rapid/pressReleasesAction.do?reference=BEI/11/113&type=HTML (accessed 10 July 2012).
EEA (European Environment Agency) (2011), *Transport Infrastructure Investments (TERM 019)*, 16 May, available at http://www.eea.europa.eu/data-and-maps/indicators/infrastructure-investments/assessment.
Esfahani, H.S. and M.T. Ramírez (2003), 'Institutions, infrastructure, and economic growth', *Journal of Development Economics,* **70**(2), 443–77.
Evenhuis, E. and R. Vickerman (2010), 'Transport pricing and public–private partnerships in theory: issues and suggestions', *Research in Transportation Economics*, **30**, 6–14.
Feitelson, E. (2002), 'Introducing environmental equity dimensions into the sustainable transport discourse: issues and pitfalls', *Transportation Research Part D: Transport and Environment*, **7**(2), 99–118.
Flyvbjerg, B., M.K. Skamris Holm and S.L. Buhl (2003), 'How common and how large are cost overruns in transport infrastructure projects?', *Transport Reviews*, **23**(1), 71–88.
Flyvbjerg, B., M.K. Skamris Holm and S.L. Buhl (2004), 'What causes cost overruns in transport infrastructure projects?', *Transport Reviews*, **24**(1), 3–18.
FRA (Federal Railroad Administration) (2009), *High-speed Intercity Passenger Rail Program*, available at http://www.fra.dot.gov/rpd/passenger/2243.shtml (accessed December 2011).
Gamez, E.A. and A. Touran (2010), 'A quantitative analysis of the performance of transportation projects in developing countries', *Transport Reviews*, **30**(3), 361–87.
Garcia-Milà, T., T. McGuire and R. Porter (1996), 'The effect of public capital in state-level production functions reconsidered', *Review of Economics and Statistics*, **78**(1), 177–80.
GEF (2012), *Behind the Numbers*, available at http://www.thegef.org/gef/sites/thegef.org/files/publication/GEF_Behind_the_Numbers_CRA.pdf (accessed April 2012).

Gibbons, S. and H. Overman (2009), *Productivity in Transport Evaluation Studies*, London School of Economics and Political Science, Commissioned by the UK Department of Transport, available at http://assets.dft.gov.uk/publications/pgr-evaluation-evaluationguidance-evalprodimpacts/productivity-in-transport.pdf (accessed March 2011).

Goetz, A. (2011), 'The global economic crisis, investment in transport infrastructure, and economic development', in K. Button and A. Reggiani (eds), *Transportation and Economic Development Challenges*, Cheltenham, UK and Northampton, MA, USA: Edward Elgar, pp. 42–71.

Graham, D.J. (2007), 'Agglomeration, productivity and transport investment', *Journal of Transport Economics and Policy*, **41**, 317–43.

Gramlich, E.M. (1994), 'Infrastructure investment: a review essay', *Journal of Economic Literature,* **32**(3), 1176–96.

Hammami, M., J.F. Ruhashyankiko and E.B. Yehoue (1999), 'Determinants of public–private partnership in infrastructure', IMF Working Paper WP/06/99, International Monetary Fund, Washington, DC.

Helme, N. (2010), *NAMAs in Key Sectors: The Way Forward*, Nordic Minister Side Event, 1 December, Mexico.

Hensher, D. and K. Button (2003), *Handbook of Transport and the Environment*, Handbooks in Transport, vol.4, Amsterdam: Elsevier.

Hickman, R., O. Ashiru and D. Banister (2010), 'Transport and climate change: simulating the options for carbon reduction in London', *Transport Policy*, **17**(2), 110–25.

High-level Panel on the CDM Policy Dialogue (2012), *Climate Change, Carbon Markets and the CDM: A Call to Action*, Luxembourg: CDM Policy Dialogue, available at http://www.cdmpolicydialogue.org/report/rpt110912.pdf (accessed 24 September 2012).

HM Treasury (2011), *UK National Infrastructure Plan*, available at http://cdn.hm-treasury.gov.uk/national_infrastructure_plan291111.pdf (accessed January 2012).

ITF (International Transport Forum) (2008), *Transport Infrastructure Investments: Options for Efficiency*, Paris: OECD.

Kahn, E.M. and D.M. Levinson (2011), 'Fix it first, expand it second, reward it third: a new strategy for America's highways', The Hamilton Project, Discussion Paper 2011-03.

Kritzinger, S., P. Beullens, J. Limbers et al. (2011), *Deliverable 2 of LogMan Project; Logistics & Manufacturing Trends and Sustainable Transport – Report on External Factors*, Funded by the European Commission 7th RTD Programme, Basle, Switzerland.

Krugman, P. (1998), *Development, Geography, and Economic Theory*, Harvard, MA: MIT Press, Chapter 1.

Lakshmanan, T.R. (2011), 'The broader economic consequences of transport infrastructure investments', *Journal of Transport Geography*, **19**, 1–12.

McQuaid, R.W. and W. Sherrer (2010), 'Changing reasons for public–private partnerships', *Public Money and Management*, **30**(1), 27–34.

New York State Press Office (2012), *Governor Cuomo Announces Accelerated Road and Bridge Projects on Long Island as Part of NY Works Program*, available at http://www.governor.ny.gov/press/04032012-roads-bridges-long-island (accessed April 2012).

Potter, S. (2008), 'Purchase, circulation and fuel taxation', in S. Ison and T. Rye (eds), *The Implementation and Effectiveness of Transport Demand Management Measures*, Hampshire: Ashgate Publishing, pp. 13–27.

Rodrigue, J.P. (2012), 'The geography of transport systems', available at http://people.hofstra.edu/geotrans/eng/ch7en/appl7en/actorstransportfinance.html (accessed 10 July 2012).

SACTRA (1999), *Transport and the Economy*, August, London: The Stationery Office.

Sakamoto, K., H. Dalkmann and D. Palmer (2010), 'A paradigm shift towards low-carbon transport financing the vision', ITDP Working Paper, Institute for Transportation and Development Policy, New York.

Sanchez, S. (2008), 'Finance for sustainable urban transport', in K.H. Olsen and J. Fenhann (eds), *A Reformed CDM: Including New Mechanisms for Sustainable Development*, Roskilde: UNEP.

Santos, G., D. Newberry and P. Rojey (2000), 'The environmental benefits of road pricing', Cambridge University Applied Economics Working Paper, available at http://www.econ.cam.ac.uk/dae/repec/cam/pdf/wp0020.pdf (accessed 9 May 2011).

Shaoul, J., A. Stafford and P. Stapleton (2006), 'Highway robbery? A financial analysis of Design, Build, Finance and Operate (DBFO) in UK roads', *Transport Reviews*, **26**(3), 257–74.

Siemiatycki, M. (2010), 'Delivering transportation infrastructure through public–private partnerships: planning concerns', *Journal of the American Planning Association*, **76**(1), 43–58.

Small, K. (2010), 'Private provision of highways: economic issues', *Transport Reviews*, **30**(1), 11–32.

STAP (Scientific and Technical Advisory Panel) (2010), *Advancing Sustainable Low Carbon Transport through GEF*, A STAP Advisory Document, Washington, DC: UNEP.

Stead, D. and D. Banister (2006), 'Decoupling transport growth and economic growth in Europe', in B. Jourquin, P. Rietveld and K. Westin (eds), *Towards Better Performing Transport Networks*, Abingdon: Routledge, pp. 136–56.

Stiglitz, J.E., A. Sen and J. Fitoussi (2009), *Report by the Commission on the Measurement of Economic Performance and Social Progress*, available at http://www.stiglitz-sen-fitoussi.fr/documents/rapport_anglais.pdf (accessed 25 February 2010).

Thatcher, M. (2008), 'Reforming national regulatory institutions: the EU and cross-national variety in European network industries', in B. Hancke and M. Rhodes (eds), *Beyond Varieties of Capitalism: Conflict, Contradictions, and Complementarities in the European Economy*, Oxford: Oxford University Press, pp. 147–72.

The Clean Air Institute (2008), *Report and Strategy to Improve the Effectiveness of CDM to Foster Sustainable Transportation*, available at http://www.transport2012.org/bridging/ressources/files/1/102,SanchezCDMcarbonReportFinal-12.1.09.pdf (accessed 10 July 2012).

Torrisi, G. (2009), 'Public infrastructure: definition, classification and measurement issues', *Economics, Management, and Financial Markets*, **4**(3), 100–24.

UNEP (United Nations Environment Programme) (2011), *Cleaner More Efficient Vehicles*, available at http://www.unep.org/transport/gfei/autotool/case_studies/europe/cs_eu_0.asp (accessed 10 July 2012).

UNEP (United Nations Environment Programme) (2012), *CDM Projects Grouped in Types*, available at http://cdmpipeline.org/cdm-projects-type.htm#4 (accessed April 2012).

Van der Bergh, J.C.J.M, E.S. van Leeuwen, F.H. Oosterhuis, P. Rietveld and E.T. Verhoef (2007), 'Social learning by doing in sustainable transport innovations: ex-post analysis of common factors behind successes and failures', *Research Policy*, **36**(2), 247–59.

World Bank (2006), *Promoting Global Environmental Priorities in the Urban Transport Sector*, The Environment Department, available at http://www.unep.org/PDF/WB_Sustainable_Transport_report.pdf (accessed 10 July 2012).

Wright, L. and L. Fulton (2005), 'Climate change mitigation and transport in developing nations', *Transport Reviews*, **25**(6), 691–717.

# 10. Supply chains

**Jian Liu**

## INTRODUCTION

Transport activities are usually divided into passenger and freight transport. Freight transport and associated logistics and supply chains constitute an important component of the modern transport system and have been regarded as fundamental elements for global economic growth. With the extension of globalization, freight transport is expected to be profoundly impacted and shifted by the changing economic demand. This chapter discusses some key issues within the freight transport and supply chains sector, including a brief introduction to global supply chain development history, the relationship between a supply chain and the modern economy, typical supply chain initiatives that affect the cost efficiency and energy intensity of freight transport operations, future freight transport development and their implications on low carbon mobility.

A supply chain should encompass all the activities involved in the planning and management of a transport system required for the production and supply of a good or service to consumers, and it consists of a network that includes people, technology, activities, information and resources. A typical supply chain begins with the recovery of natural resources, followed by the human extraction of raw material, and includes several production stages (from component construction, assembly, to merging) and moving to storage facilities, and finally reaching the consumer. A supply chain is a dynamic system and involves the constant flow of information, product and funds between the different stages.

## SUPPLY CHAIN STRUCTURE

A supply chain can be referred to as an integrated system that synchronizes a series of inter-related business processes covering (Min and Zhou, 2002; Figure 10.1):

1. Acquiring raw materials and parts.
2. Transforming raw materials and parts into finished products.
3. Adding value to products.
4. Distributing and promoting products to retailers or customers.
5. Exchanging information among business entities (for example, suppliers, manufacturers, distributors, third party logistics providers and retailers).

The objective of the supply chain operation is to link various nodes within a supply chain network and enhance the efficiency of logistics. Typically, a supply chain is comprised of two main business processes:

- material management (inbound logistics);
- physical distribution (outbound logistics).

Material management (inbound logistics) refers to the acquisition and storage of raw materials, parts and supplies. According to Johnson and Malucci (1999), material management contains the cycle of material flow from the purchase and internal control of materials to the planning and control of work-in-process, to the warehousing, transport and distribution of final products. On the other hand, physical distribution (outbound logistics) covers all outbound logistics activities related to providing customer service, namely, order receipt and processing, inventory deployment, storage and handling, outbound transport, consolidation, pricing, promotional support, returned product handling and life cycle support (Bowersox and Closs, 1996). Along with the process of the material management and the physical distribution, different stakeholders (comprised of various suppliers, manufacturers, distributors, third party logistics providers, retailers and customers) could be involved and form multiple business networks and relationships in the supply chain.

When structuring a supply chain network, there is a need to identify the partnership within the supply chain. Lambert et al. (1998) classified supply chain partners into two types: primary and supporting partners. The primary partners are those autonomous channel captains or strategic business units that perform operational and/or managerial activities designed to create a specific product or service for a particular customer or market. These primary partners can be manufacturers such as Dell or mass merchants such as Wal-Mart. In contrast, supporting partners simply provide resources (including assets, software and real estate property), knowledge and utility (water, electricity and gas) for the supply chain. These supporting partners can be transport carriers, consulting companies and third party logistics providers and information technology providers. However,

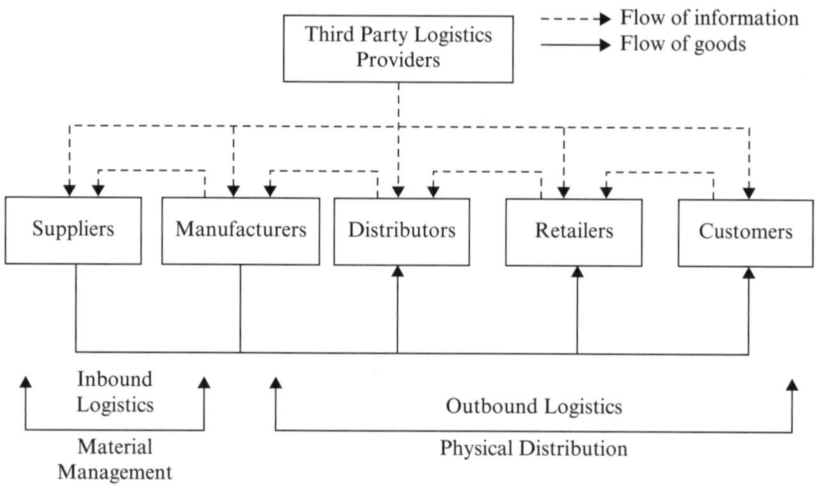


*Source:* Min and Zhou (2002).

*Figure 10.1 Supply chain structure*

the classification is not exclusive and a primary partner can also be a supportive partner of the supply chain. Within a supply chain the furthest upstream members usually represent supporting partners, while the furthest downstream represent the end of the supply chain where no further value is added and the product or service is consumed.

## SUPPLY CHAIN MANAGEMENT

Supply chain management is defined as 'the integration of key business processes from end-users through original suppliers that provide products, services, and information and add value for customers and other stakeholders' (Cooper et al., 1997). Effective management of a supply chain has been recognized as a key factor in differentiating product and service offerings and gaining competitive advantage for firms (Christopher, 1998). Supply chain management consists of a set of synchronized decisions and activities utilized to efficiently integrate suppliers, manufacturers, warehouses, transporters, retailers and customers so that the right product or service is distributed at the right quantities, to the right locations and at the right time in order to minimize system-wide costs while satisfying customer service level requirements.

A variety of supply chain management techniques have been proposed to improve the delivery efficiency and to reduce energy intensity and thus emissions in the logistics industry. They can be generally summarized as consisting of improvements to supply chain operation efficiency with a minimal investment in infrastructure and environmental impact (storehouse, vehicle fleet and emissions); and development of a 'seamless' transport system to minimize the idle and transfer duration of a low stock level in each segment. Some specific supply chain management techniques to improve efficiency and reduce emissions are described below. Measures to increase supply chain efficiency can reduce supply chain costs and related emissions, but can potentially result in increased demand (since costs are lower) and thus in the longer term to more emissions, a rebound effect.

### Establishing a Monitoring Mechanism

An efficient supply chain relies on a monitoring mechanism to measure both costs and associated emissions. For instance, the 'Carbon Footprint Metrics' strategy has already been adopted to evaluate the life cycle carbon intensity of a product and illustrate the carbon emission variation along different stages (Wang, 2010). The establishment of a monitoring mechanism system is also a crucial requirement to assess the consumption of vehicle fuels, used metals, tyres, batteries, vehicle components and all materials that are involved in the supply chain. According to the United States–Asia Environmental Partnership's (USAEP) estimation, recycling every tonne of solid fibre can save 463 gallons of crude oil, 24 gallons of gasoline, 4100 kWh of electricity, 7000 gallons of water and 3 square metres of storage space (USAEP, 2001).

### Demand and Supply Optimization

Optimizing the match between demand and supply has been widely applied in logistics. It usually involves more precise order prediction, collective processing and optimized storage management. At present, some developed countries have introduced 'lean distribution' indicators for distribution centre management to minimize the waste by improving space utility. Some storage centres have applied green architecture technologies and have been using renewable energy, such as photovoltaic panels and waste water recycling infrastructure (Dong, 2010). Such activities can reduce operation costs and emissions of carbon and other pollutants, also contributing to a reduction in air pollution.

Optimizing supply chain networks' nodal points, hierarchy and interrelated transport flows has resulted in significant reductions in both cost and carbon. Accenture Research has shown that restructuring the network gave an 11 per cent cost reduction (Doherty and Hoyle, 2009). There is a large potential for further improvements. For example, 24 per cent of goods vehicle tonne-km in the European Union (EU) are running empty and when carrying a load, vehicles are typically only 57 per cent loaded as a percentage of maximum gross weight (WEF, 2009). There is subsequently a large cost reduction potential by improving economic transaction sizes in freight movements, though empty operation is always likely to be a feature of freight transport and would be difficult to eliminate completely.

**Logistics Management**

Freight transport can be improved through optimizing routes, minimizing no-load time (empty running), improving freight consolidation and applying new energy vehicles. In addition, by improved transport simulation and utilizing global positioning systems (GPS), the delivery efficiency can be largely increased, particularly in urban areas. A supply chain driven by factors such as lead times, deadlines and booking windows requires a fast delivery. The speed of a supply chain could be improved by adopting new vehicle/ship and load-fill improvements with reduced time windows. However, fast transport usually means high energy consumption and emissions across the supply chain. The balance between freight transport efficiency and its carbon footprint has become a key issue in the modern freight transport and logistics industry.

**Multi-mode Delivery**

Among different transport modes road freight transport is still largely preferred compared to air and rail, and even when air and rail are used road transport is still often used for the ‘last mile’. This is mainly due to the high flexibility and accessibility it offers to customers compared with other modes, and, therefore, it plays a vital role in the freight transport industry. Particularly for short distances and small volume delivery, road transport probably provides the highest efficiency with the lowest cost. Yet, the delivery capacity of road transport is relatively small, which leads to a high energy consumption along the supply chain. There are large differences in carbon emissions between different freight transport modes when measured in emission per tonne-km shipped (Figure 10.2). Among different transport modes the most carbon efficient modes are rail and ocean freight. The carbon intensity of both modes is quoted by the UK Department for the Environment, Food and Rural Affairs (DEFRA) as being around one-sixth of road freight – or one-hundredth that of air

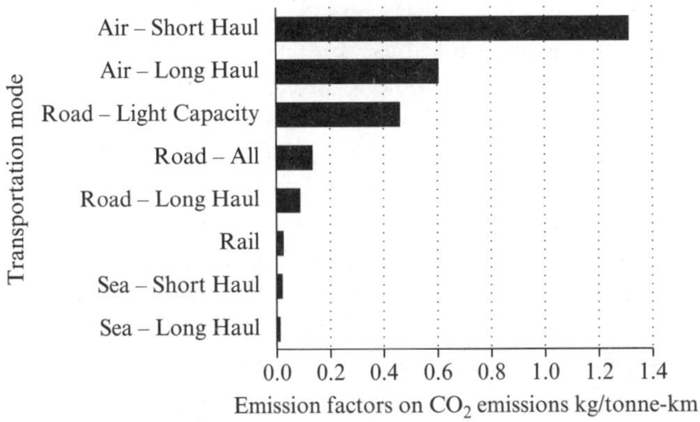


*Note:* It is assumed that each mode is operating fully laden. Emission factors will significantly differ when different load factors are applied.

*Source:* World Economic Forum (2009).

*Figure 10.2 Emission factors of transport modes*

freight (DEFRA, 2008). Therefore, a significant reduction in emissions can be achieved by switching short haul air to road transport and assigning long distance road freight to rail or waterways. Yet, to encourage and support such a shift, corresponding policy incentives and political support are required to improve the competitiveness of the modes that are environmentally more efficient. It should be noted, however, that small-volume goods shifted, for example, from road to rail or waterways could lower the freight efficiency of these modes (McKinnon et al., 2010).

A supply chain based on container shipping and the use of several modes of transport can largely reduce both the costs and the time for the operator and lower the transport energy consumption and emissions. In the US to deliver the same volume of cargo, the multi-mode delivery of integrated railway and highway transport with container shipping can reduce carbon emissions by 60–70 per cent compared with highway transport (BNSF, 2008).

**Home Delivery**

It can be assumed that collective home delivery by retailers to customers is more efficient than having consumers drive to stores to make their purchases. The last decade has seen a surging growth of electronic commerce, which has fundamentally changed the freight transport industry. Online shopping companies, such as eBay and Amazon, provide a new shopping platform for customers, and these outlets have become a challenging competitor to conventional retailers. Internet shopping also impacts on consumers' shopping style, and consequently changes the freight transport and supply chain industry. With the fast development of e-business and information and communication technologies (ICT), home delivery has been substantially increased in certain segments of the retail market. There also remains a significant potential in the market for growth, with market volumes growing by around 20 per cent per annum in several Western economies. In addition to providing high delivery efficiency, home delivery also brings environmental benefit. According to the World Economic Forum's (WEF) analysis, home delivery is shown to be around four times more efficient in terms of carbon emissions, if it is assumed that the home delivery service eliminates the shopping travel journey entirely (Doherty and Hoyle, 2009). Moreover, in addition to the energy saving and environmental benefits, home delivery can also provide significant commercial opportunities for retailers and promote profits (Williams and Tagami, 2002).

## DEVELOPMENT OF GLOBAL SUPPLY CHAINS

With the fast development of the global economy and international trade, the logistics system has developed with immense and interweaving supply chain networks across the world (WTO and World Bank, 2011; Figure 10.3). In recent years, as one of the major worldwide economic activities in modern society, the logistics industry, along with expanding worldwide supply chain networks, has become one of the most important issues in freight transport studies. The development of modern supply chains is stimulating and stimulated by economic development. Robust economic growth requires a well-established supply chain system that provides the platform for international trade and business, and economic growth in turn further stimulates supply chains development (see also Chapter 9).

In the early twentieth century transporting goods was still relatively time-consuming and labour-intensive and long distance transportation of products was relatively inefficient. High shipping costs offered shelter to high-cost producers whose significant advantage was proximity to their customers. With custom duties and time delays, high transport cost was a significant barrier to the globalization of the world economy. The situation, however, started changing from the 1950s when the efficiency of freight transport started to improve through a series of technological improvements such as the application of ICT, supply chain management techniques and innovative logistics designs. In particular, the design of the

*Note:* GDP – gross domestic product.

*Source:* WTO and World Bank (2011).

*Figure 10.3 World merchandise trade, 1960–2010*

container, arguably the main supply chain innovation of the last century, largely reduced the cost of long distance freight transport, and by doing so changed the shape of the world's economy (Levinson, 2006). Supply chain innovations have given multinational manufacturers and companies the ability to locate plants in different countries, and to transform themselves into international manufacturers, integrating once isolated factories into a network so that they could choose the cheapest location to make a particular item and reduce their overall cost. In the 1950s the world was full of small manufacturers that concentrated on the local market, but by the end of the twentieth century, purely local markets had largely shrunk, and the improvement in supply chains has stimulated global trade, offering more choices for customers.

In recent decades international trade growth has outpaced population growth. Although the Asian economic crisis during 1997 and 1998 and the recent global economic recession decelerated growth rates, economic growth particularly in developing countries, such as China, India and Brazil, continued to drive the development of international trade (WTO, 2010). In the US the annual imported goods in 2002 were four times greater than in 1972, generating nearly 3 per cent of the entire economy (Levinson, 2006). The expected continued growth in global trade will continue to promote the growth of supply chains.

*Table 10.1 Value of world export in billion US dollars by countries, 1993–2010*

| | 1993 | 2010 |
|---|---|---|
| World (billion dollars) | 3676 | 14851 |
| World (percentage) | 100.0 | 100.0 |
| North America | 18.0 | 13.2 |
| US | 12.6 | 8.6 |
| Canada | 3.9 | 2.6 |
| Mexico | 1.4 | 2.0 |
| South and Central America | 3.0 | 3.9 |
| Brazil | 1.0 | 1.4 |
| Argentina | 0.4 | 0.5 |
| Europe | 45.4 | 37.9 |
| Germany[a] | 10.3 | 8.5 |
| France | 6.0 | 3.5 |
| Italy | 4.6 | 3.0 |
| UK | 4.9 | 2.7 |
| Commonwealth of Independent States (CIS)[b] | 1.5 | 4.0 |
| Africa | 2.5 | 3.4 |
| South Africa[c] | 0.7 | 0.6 |
| Middle East | 3.5 | 6.0 |
| Asia | 26.1 | 31.6 |
| China | 2.5 | 10.6 |
| Japan | 9.9 | 5.2 |
| India | 0.6 | 1.5 |
| Australia and New Zealand | 1.4 | 1.6 |
| Six East Asian Traders | 9.7 | 10.1 |
| Memorandum item: | | |
| EU[d] | 37.4 | 34.7 |
| USSR, Former | – | – |
| GATT/WTO Members[e] | 89.3 | 94.2 |

*Notes:*

a. Figures refer to the Federal Republic of Germany from 1948 to 1983.

b. Figures are significantly affected by including the mutual trade flows of the Baltic States and the CIS between 1993 and 2003.

c. Beginning with 1998, figures refer to South Africa only and no longer to the Southern African Customs Union.

d. Figures refer to the EEC(6) in 1963, EC(9) in 1973, EC(10) in 1983, EU(12) in 1993, EU(25) in 2003 and EU(27) in 2010.

e. Membership as of the year stated. WTO – World Trade Organization; GATT – General Agreement on Tariffs and Trade.

Between 1973 and 1983 and between 1993 and 2003 export shares were significantly influenced by oil price developments.

*Source:* WTO (2010).

Along with the development of water and road freight transport, growth in air freight transport has occurred after the Second World War. Compared with other freight transport modes, air freight offers a considerably faster delivery. However, due to its high cost, air freight transport remained a minor portion of overall freight transport, when measured in volume of goods, and its development fell far behind that of passenger air transport development until the 1970s. Since the 1980s, however, with the introduction of exclusive air freight delivery services by Federal Express and UPS, the large potential of air freight has started to be realized by the logistics industry. With the growing demand for reducing transport time and progress in ICT, the growth of air freight transport is expected to continue in the future (Bowen and Leinbach, 2006). Boeing (2011) forecasts that air cargo traffic (based on revenue tonne-km) will grow at an average annual rate of 5.6 per cent between 2011 and 2030, mainly due to growing world trade and increasing demand for the transport of perishable and time-sensitive commodities.

Over the last two decades global supply chain development has been significantly influenced by the transfer of manufacturing industry from developed countries to developing countries, particularly to Eastern Asian countries. World exports are growing faster than the world GDP growth and the trend is consistent across regions and industries (WTO, 2007). The manufacturing, distribution, sourcing of materials, invoicing and returns have all been largely impacted by the increased integration of global consumers and suppliers. There is a remarkable trend of cheaper import demand from growing economies, such as China, India, the Middle East and Russia. Six of the top ten destinations for retail investments and three of the top ten destinations for heavy manufacturing were not on the list in 2003 (A.T. Kearney, Global Retail Development Index, 2007). As a result, the average length of freight transport distances has been growing longer and longer. In addition, logistics companies have been forced to follow customers who have employed 'offshoring' strategies – the relocation of production to low-wage countries.

## SUPPLY CHAINS AND THE ECONOMY

Currently, worldwide supply chains have a significant impact on the world's economy. Such impacts were clearly seen after the earthquake in Japan in 2011. Some global automotive, building infrastructure and electronic industries have substantially suffered from suspension of production in Japanese factories (Lee and Pierson, 2011). While the Japanese government worked to fix the nuclear reactors and restore order, it required the major manufacturers to shut down operations until further notice so that all energy could be used for emergency purposes. This had major

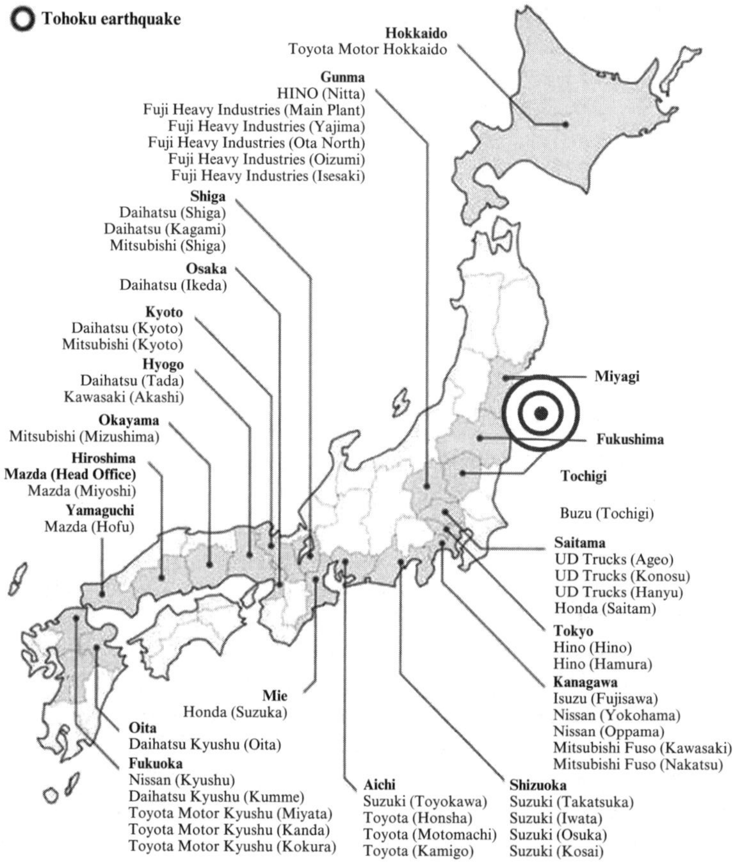


*Source:* 21cn.com (2011).

*Figure 10.4 Tohoku earthquake and Japanese factories distribution*

implications for global supply chains depending on the closure duration of the factories. Japanese-made equipment and materials play a key production and market role in many of the region's main industries, from automobiles to semiconductors. Although many Japanese car manufacturers, including Honda, Toyota and Nissan, did not see significant damage to production facilities, the damage to Japan's supply chain infrastructure, which links their factories across the country, limited their ability to move cars and products throughout the world (Figure 10.4). In a similar case, the Icelandic volcano ash eruption that occurred in 2010 not only caused a major disruption to European travellers trying to get home, but also had a knock-on effect on the global trading market: car-maker BMW suspended production at three of its plants in Germany because of interruptions in the supply of parts (Beattie, 2012); in Japan Nissan suspended production lines, while Honda announced a partial halt to production; in China blocked shipments of goods stacked up, while South Korea was stuck with hundreds of thousands of mobile phones; and in Bangladesh heaps of clothing bound for Europe piled up. The loss of air freight due to the volcano ash has cost the economy more than £1.6 billion (Densley, 2010).

These incidents reflect the fact that accelerated globalization and international trade in recent decades have largely established the world's supply chain system and freight transport, which has played a vital role in the global economy. Though there is significant uncertainty in exact figures, it has been widely accepted that the future global economic growth largely depends on a reliable and robust supply chain system. The development of the logistics industry can therefore be further promoted through a wider adoption of available technologies, and advanced management techniques, leveraging new commercial relationships and developing new business strategies.

## SUPPLY CHAIN AND FREIGHT TRANSPORT'S SUSTAINABILITY

Although freight transport is currently dominated by rail and maritime shipping, which are the two most energy efficient modes, when measured in tonne-km, the volume of freight transport is forecast to further increase due to economic growth and, for example, the development of e-commerce and related businesses, and this will increase energy use and emissions from freight transport. Since the supply chain plays an important role in modern society, the efficiency of the supply chain is critical for sustainable development. The application of appropriate supply chain management techniques will not only improve the efficiency of the transport flow but also lower the energy intensity and carbon emissions embedded within the flow. Nowadays, of humanity's 50 000 mega-tonnes of annual $CO_2$ greenhouse gas (GHG) emissions, around 2800 mega-tonnes can be assigned to logistics and transport activities (World Economic Forum, 2009). With the increasing share of transport in the economy, particularly in developing countries, the development of a low-cost and efficient supply chain requires a particular focus. An efficient supply chain could not only mitigate a country's reliance on imported oil but also reduce the associated carbon emissions. For example, although the Chinese government has made the commitment to cut 40–45 per cent carbon emissions per GDP unit by 2020 compared with the 2005 level (NDRC, 2010), China is still facing prominent challenges in carbon reduction especially in the freight transport sector. Since China launched its economic reform process in 1978, the average GDP growth has averaged 9.5 per cent annually. Within a very short period of time, from 2002 to 2007, China's $CO_2$ emissions have almost doubled and China currently is the world's biggest emitter of $CO_2$ (Gregg et al., 2008). China faces increasing international pressure to curb its $CO_2$ emissions growth, but argues that 'the ramifications of limiting the development of developing countries would be even more serious than those from climate change' (Wu, 2007). With China's processing/manufacturing based industry extending both upstream (resources, raw materials and energy recovery industries) and downstream of supply chain (product distribution and sales), a highly efficient supply chain becomes one of the crucial measures to achieve the carbon reduction target.

A remarkable effort in promoting a sustainable supply chain system is the application of life cycle assessment or analysis (LCA) that can investigate the energy use and emissions along the whole of the supply chain. LCA is intended to provide a thorough analysis of each step of a product life cycle by compiling an inventory of relevant inputs and outputs of the system, evaluating the potential economic and environmental impacts associated with these inputs and outputs and interpreting the results in relation to the objects of the study (ISO, 1997). A full LCA involves all the activities associated with a product or service. This is often referred to as the supply chain of the product, including: acquisition of raw materials; manufacturing and processing; storage, materials handling and logistics management; freight transport; use and maintenance; and recycling/reuse or waste management. LCA is an effective tool to separate and compare different logistics stages across the supply chain. It can also be applied to identify the potential sourcing and distribution centre locations, the optimum number of storage points, suitable transport modes and vehicle types (Browne et al., 2005).

Managing the carbon life cycle of a product requires an optimization of the logistic efficiency throughout the product's carbon footprint all the way to the end consumer. For example, the carbon life cycle of a can of cola must account for every step in the product supply chain including the production, use and disposal processes (Figure 10.5).

LCA analysis helps us to understand the reasons why costs are generated across the economy. Processes, and their costs, do not occur in isolation but are always part of the supply chain of different products or services. At the economy-wide level, it is possible to take every cost source and allocate it to the supply chain of a different product or service. The result is to show that all the costs generated in an economy exist to deliver products

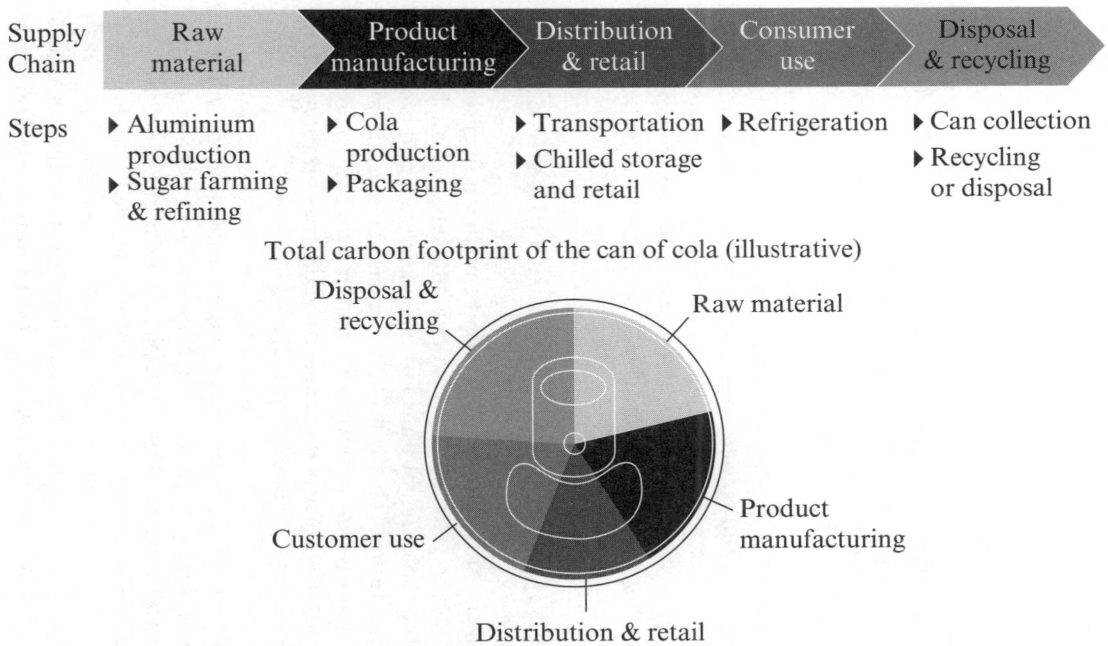


*Source:* Carbon Trust (2006).

*Figure 10.5 The supply chain and carbon life cycle of a can of cola*

and services to meet the needs of the end consumer. At the individual product level, this supply chain approach has the potential for significant cost reduction opportunities and large financial benefits by reducing the overall cost of the product. It can help to explain the costs across different supply chains and allow areas to be prioritized where further reductions in costs can be achieved. It can ultimately help us to make better informed decisions in product manufacturing, purchasing, distribution and product development by considering the costs and liabilities that exist whenever unnecessary costs are generated. A number of other strategies have also been proposed (WEF, 2009) to improve the supply chain efficiency in terms of time, cost and carbon emissions abatement potential and feasibility. Some typical strategies have been introduced above.

It should be noted that with the expansion of the global supply chain, the developed world is 'outsourcing' a large amount of emissions to developing countries. A significant proportion of carbon emissions associated with the consumption of goods and services in industrialized countries are emitted in other countries. Due to the impediment to limit carbon emissions in developing countries, some portion of the carbon emissions responsibilities should be allocated to final consumers (Davis and Caldeira, 2010).

## FUTURE FREIGHT TRANSPORT DEVELOPMENT

Future development of freight transport will be mainly influenced by the following. First, due to growth in international markets, a clear shift from national to international freight transport has taken place and is expected to continue in the future. This will likely increase congestion along the trade corridors (and ports, airports and border crossings) and impose challenges in terms of infrastructure capacity and harmonization of trade and regulatory policies (Site and Salucci, 2009).

Second, with the shift from manufacturing to a service economy and the accompanied growth of e-commerce and e-business, there will be an increasing volume of light, high-value shipment in customer-oriented door-to-door truck services. For example, General Electric uses the internet to schedule shipments out of centrally located warehouses to allow the company to save delivery time and cost (Lancioni et al., 2000). Major retailers such as Dell and Tesco sell products online as well as through their traditional distribution networks of warehouses and stores. Providers such as Apple, Amazon and eBay even allocate the majority of their products/services on their own web stores. On the other hand, although e-business could induce a substantial growth in freight transport in the future, it could also provide measures to improve the supply chain capacity. With the help of advanced ICT, more supply chain management techniques are expected to be applied to reduce the freight transport cost as well as improve the sustainability of the logistics industry. In addition, logistics security will receive greater attention in future supply chains. The implementation of new technologies, such as radio frequency identification (RFID) tags, is one example. Also expected is that the demand of multi-modal transport services will increase, but will need to be reliable and predictable to gain market share (Site and Salucci, 2009).

Finally, environmental impacts of freight transport including emissions, congestion and accidents will become even more important as the volume of freight transport increases. Expected increases in the cost of fuel will continue to encourage the development of low carbon supply chains.

## CONCLUSION

This chapter has reviewed the history of worldwide supply chain development in the context of the growing global economy and international trades. It has investigated the supply chain management concepts, theories and strategies, with attention given to LCA and carbon footprint methods. It has shown that the development of a modern supply chain relies on a series of coordinated strategies, including introducing advanced supply chain management techniques, improving supply and demand prediction and balancing, optimizing the transport network and adopting more efficient freight transport modes. Moreover, although freight transport efficiency has been improved through improved logistics and supply chain management over recent decades, some other challenges can also be observed. For example, the introduction of the Emission Trading Scheme (ETS) has reduced manufacturing emissions, but increased transport emissions (which are largely excluded from the ETS). The global division of the supply chain and labour is also inducing more freight transport, resulting in greater energy consumption and emissions. In addition, the chapter has shown that many traditional logistics and freight transport management techniques are also effective in establishing a low carbon supply chain, and new strategies to further improve efficiency will not only reduce operating costs and cut GHG emissions, but also provide new opportunities to formulate new models for a sustainable economy. Yet, there is a risk that efficiency measures that reduce costs along the supply chain, and also emissions, will in the longer term result in more freight transport activity and even longer supply chains, thus offsetting any emission reduction from increased efficiency by an increase in the absolute volume of freight transport (for further discussion of these issues, see Part II of this book).

# REFERENCES

21cn.com (2011), *Japan's Earthquake Impacts on Car Makers – Affected Factories*, available at http://auto.21cn.com/news/hangye/2011/03/14/8170987.shtml (accessed 2 May 2011).
A.T. Kearney, Global Retail Development Index (2007), http://www.atkearney.com/index.php/Publications/the-2007-global-retail-development-index.html (accessed 25 April 2012).
Beattie, A. (2012), 'Poor disaster plans threaten global economy', *Financial Times*, available at http://www.ft.com/cms/s/0/7ecbc672-37b2-11e1-897b-00144feabdc0.html#axzz1kUXvTlae (accessed 8 January 2012).
BNSF (Burlington Northern Santa Fe) (2008), *BNSF Railway Carbon Estimator*, available at http://domino.bnsf.com/website/carbon.nsf/ce?open (accessed 20 September 2011).
Boeing (2011), *Current Market Outlook 2011–2030*, Chicago, IL: Boeing.
Bowen, J. and T. Leinbach (2006), 'Competitive advantage in global production networks: air freight services and the electronics industry in SE Asia', *Economic Geography*, **82**(2), 147–66.
Bowersox, D.J. and D.J. Closs (1996), *Logistical Management: The Integrated Supply Chain Process*, New York: McGraw-Hill.
Browne, M., C. Rizet, S. Anderson, J. Allen and B. Keita (2005), 'Life cycle assessment in the supply chain: a review and case study', *Transport Reviews*, **25**(6), 761–82.
Carbon Trust (2006), 'Carbon footprints in the supply chain: the next step for business', Carbon Trust Company, London.
Christopher, M. (1998), *Logistics and Supply Chain Management: Strategies for Reducing Costs and Improving Services*, 2nd edn, London: Financial Times/Pitman Publishing.
Cooper, M.C., D.M. Lambert and J.D. Pagh (1997), 'Supply chain management: more than a new name for logistics', *International Journal of Logistics Management*, **8**(1), 1–13.
Davis, S. and K. Caldeira (2010), 'Consumption-based accounting of $CO_2$ emissions', Department of Global Ecology, Carnegie Institute of Washington, Stanford, California.
DEFRA (Department for the Environment, Food and Rural Affairs) (2008), *DEFRA Emissions Factors*, available at http://naei.defra.gov.uk/emissions/selection.php (accessed 9 May 2011).
Delivering Results (n.d.), available at http://www.con-way.com/resources/BRT_ConwaySustainabilityUpdate2008_.pdf (accessed 10 May 2011).
Densley, R. (2010), 'Volcanic ash cloud causing disruption to supply chain', *NG Supply and Logistics*, available at http://www.ngsupplyemea.com/news/Volcanic-ash-cloud-causing-disruption-to-supply-chain/ (accessed 28 January 2012).
Doherty, S. and S. Hoyle (2009), 'Supply chain decarbonisation: the role of logistics and transport in reducing supply chain carbon emissions', World Economic Forum, available at http://www.idekompasset.dk/files/Rapporter/SupplyChainDecarbonization.pdf (accessed 9 May 2011).
Dong, Q.L. (2010), 'Low carbon logistics theories and strategies', *Science and Technology Progress and Policy*, **27**(22), 22.
Gregg, J.S., R.J. Andres and G. Marland (2008), 'China: emissions pattern of the

world leader in $CO_2$ emissions from fossil fuel consumption and cement production', *Geophysical Research Letters*, **35**, L08806.
ISO (1997), *The European Standard EN ISO 14040: Environmental Management – Life Cycle Assessment – Principles and Framework*, London: British Standards Institution.
Johnson, D.A. and L. Malcucci (1999), 'Shift to supply chain reflects more strategic approach', *APICS – The Performance Advantage*, October, 28–31.
Lambert, D.M., M.C. Cooper and J.D. Pagh (1998), 'Supply chain management: implementation issues and research opportunities', *International Journal of Logistics Management,* **9**(2), 1–19.
Lancioni, R.A., M.F. Smith and T.A. Oliva (2000), 'The role of the internet in supply chain management', *Industrial Marketing Management*, **29**(1), 45–56.
Lee, D. and D. Pierson (2011), 'Disaster in Japan exposes supply chain flaw', *Los Angeles Times*, available at http://articles.latimes.com/2011/apr/06/business/la-fi-quake-supply-chain-20110406 (accessed 11 January 2012).
Levinson, M. (2006), *The Box*, Princeton, NJ: Princeton University Press.
McKinnon, A., S. Cullinane, M. Browne and A. Whiteing (2010), *Green Logistics: Improving the Environmental Sustainability of Logistics*, The Charted Institute of Logistics and Transport, London.
Min, K. and G. Zhou (2002), 'Supply chain modeling: past, present and future', *Journal of Computers and Industrial Engineering*, **43**, 231–49.
NDRC (National Development and Reform Committee) (2010), 'China will cut 40 per cent–45 per cent carbon–GDP intensity by 2020', available at http://www.gov.cn/wszb/zhibo409/content_1712489.htm (accessed 2 October 2010).
Site, P.D. and M.V. Salucci (2009), *Thematic Research Summary – Freight Transport*, DG Energy and Transport, European Commission, Transport Research Knowledge Centre, available at http://www.transport-research.info/Upload/Documents/201002/20100215_130009_1238_TRS_Freight_Transport.pdf (accessed 13 December 2011).
USAEP (United States–Asia Environmental Partnership) (2001), *Greening the Supply Chain*, United States–Asia Environmental Program, available at http://www.usaep.org/ctem/greening.htm (accessed 30 June 2012).
Wang, G.W. (2010), 'Low carbon logistics and green supply chain: concept, process and policies', *China Opening Herald*, **2**(149).
WEF (World Economic Forum) (2009), *CEO Climate Policy Report to G8 Leaders*, available at https://members.weforum.org/pdf/environment/CEOStatementSP.pdf (accessed 9 May 2011).
Williams, E. and T. Tagami (2002), 'Energy use in sales and distribution via e-commerce and conventional retail: a case study of the Japanese book sector', *Journal of Industrial Ecology*, **6**(2), April, 99–114.
WTO (World Trade Organization) (2007), *International Trade Statistics*, available at http://www.wto.org/english/res_e/statis_e/its2007_e/its07_toc_e.htm (accessed 25 April 2012).
WTO (2010), *International Trade Statistics 2010*, available at https://www.wto.org/english/res_e/statis_e/its2010_e/its2010_e.pdf (accessed 20 September 2011).
WTO (World Trade Organization) and World Bank (2011), *World Merchandise Trade, 1960–2010*, available at http://people.hofstra.edu/geotrans/eng/ch5en/conc5en/worldexports.htm (accessed 18 October 2011).
Wu, C. (2007), 'Action plan aims to cut gas emissions', *China Daily*.

# 11. Transport and the environment

**Andre Neves**

## INTRODUCTION

Transport has been essential in the development of the post-industrial society. Modern lifestyles and the economic growth of the last decades have relied extensively on the fact that transport enables people and goods to move fast, long distances and at affordable prices. These new patterns of mobility are, however, becoming increasingly associated with impacts on the environment. The transport sector is one of the main sectors responsible for the increase in emissions of greenhouse gases (GHGs) and the subsequent impacts on climate change. Moreover, demand for transport and travel behaviour choices together with the increase in transport infrastructure have been associated with other environmental impacts such as the increase in air pollution and noise exposure, changes in land use, fragmentation of habitats or effects on biodiversity and wildlife among others (EEA, 2009).

This chapter examines the transport sector's impact on the environment and the challenges in moving towards a more sustainable transport system and to low carbon mobility. The first section examines how and to what extent the transport sector has negative effects on the environment. Various modes of transport are considered including: road, aviation, rail and maritime transport with respect to their individual contributions to GHG emissions, air pollution, noise, and their impact on land resources and wildlife. Safety is another important environmental impact of transport, but it is not covered here since it is less relevant for the low carbon mobility debate. The second section explores how the different environmental impacts associated with transport can be reduced to facilitate a transition towards a more sustainable transport system. The main trends, strategies and individual measures that will reduce transport sector effects on the environment are identified and their impacts discussed.

# TRANSPORT IMPACTS ON THE ENVIRONMENT

The transport sector is considered a central element in the global economy and in modern lifestyles by supporting the increased mobility of passengers and freight around the world (Chapters 9 and 10). However, the growing expansion of transport activities is associated with significant direct, indirect and cumulative impacts on the environment (Rodrigue, 2009), issues that have become more important in recent years. The magnitude of these impacts varies between transport modes and it is also dependent on energy sources, life cycle costs and geographical scale. This section examines the extent of these impacts and analyses how the transport sector has become a main contributor to them.

## Transport and Climate Change

The contribution of the transport sector to overall GHG emissions and its impacts on climate change has been considered by many as unsustainable, and as one of the most challenging environmental concerns (Banister, 2011b). Simultaneously, changes in climate and the consequences for global warming are increasingly more evident, in terms of the frequency, intensity and worldwide location of climate-related events.

The energy balance of the climate system can be modified due to changes in the amount of GHGs and aerosols in the atmosphere, changes in solar radiation or changes in land surface properties. Climate change refers to a change in climate over time and was, historically, due to natural variability. However, since the Industrial Revolution and mostly during the second half of the twentieth century, the combustion of fossil fuels and change in land use associated with anthropogenic actions have significantly increased the concentrations of carbon dioxide,[1] methane and nitrous oxide in the atmosphere. Outgoing heat radiation from the Earth is now absorbed by the atmosphere and reradiated back towards the surface – this process is known as the greenhouse effect – resulting in global warming and an increase in the Earth's average temperature.

The consequences of climate change and its adverse effects on natural and human systems are wide ranging and well documented by the Intergovernmental Panel on Climate Change (IPCC). Projected impacts include melting of arctic ice; sea levels rise; frequent extreme weather events such as heat-waves, floods or droughts; loss of biodiversity; spread of invasive species; diminution of freshwater supplies; turmoil events, spread of diseases or increased mortality, among others.

The transport sector is a primary source of key energy-related GHG emissions, along with energy production and the energy industry

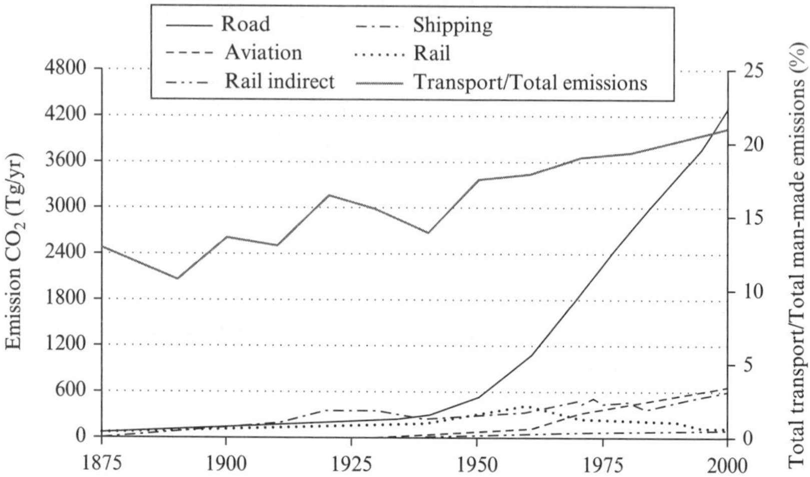


*Note:* The right axis indicates the total anthropogenic $CO_2$ emissions for the same period, excluding land use changes.

*Source:* Fuglestvedt et al. (2008). Copyright © 2008 National Academy of Sciences, USA.

*Figure 11.1 Emissions of $CO_2$ by transport mode between 1875 and 2000*

(Unruh, 2000; IPCC, 2007). The transport sector alone is responsible for nearly one-quarter of all GHG emissions in Europe (EEA, 2009) and for 22 per cent of GHG emissions worldwide (IEA, 2010). Furthermore, while emissions from other sectors are slowly stabilizing, emissions from the transport sector keep growing at a fast pace.[2] Carbon dioxide ($CO_2$) is the predominant greenhouse gas emitted within the transport sector, accounting for 98.7 per cent of its total GHG emissions (EEA, 2011a). Other GHGs include methane ($CH_4$) and nitrous oxides ($NO_x$) and although both are emitted in much smaller quantities, their impact on global warming is still significant (Forster and Ramaswamy, 2007).

The different modes of transport contribute to climate change through GHG emission in a distinctive way. Road transport is, incontestably, the largest source of GHGs; however, in the last decade aviation and maritime transport have also increased their contribution significantly (Figure 11.1).

The level of $CO_2$ emitted from passenger and freight transport can vary according to the type of fuel used, vehicle type, weight and volume of goods transported, and logistic efficiency (IEA, 2009). Table 11.1 summarizes the average $CO_2$ emission for passenger and freight transport by

*Table 11.1 Passenger and freight transport average $CO_2$ emissions per passenger and tonne-km*

| | | Passenger transport efficiency (average $CO_2$ emissions per pkm) g/pkm | Freight transport efficiency (average $CO_2$ emissions per tkm) g/tkm |
|---|---|---|---|
| Road | Car (1.1 passenger) | 95–480 | 62–110 |
| | Bus or coach | 45–80 | |
| Aviation | Air (short haul) | 77–240 | +665 |
| | Air (long haul) | 118–153 | |
| Rail | Rail normal | 45–130 | 18–35 |
| | Rail (high speed) | 80–165 | |
| Maritime transport | Sea transport | – | 2–7 |
| | Inland waterways | | 30–49 |

*Source:* Based on EEA (2008, 2009).

different modes. In this respect, aviation is often the least efficient mode of transport, while maritime transport and rail transport can be very efficient in transporting goods (2–7 g per tonne-km) and buses in transporting passengers (45–80 g per passenger-km) (EEA, 2009).

Road transport is the largest contributor to GHG emissions within the transport sector, accounting for nearly 73 per cent of total transport $CO_2$ emissions worldwide (IEA, 2010). Cars are the most popular passenger mode of transport and contribute more than 6 per cent of anthropogenic $CO_2$ emissions worldwide (Stern, 2006). The fact that distance travelled by car[3] and car ownership[4] continue to grow at an incredible pace in some countries, like China, India and others, represents a further concern in terms of GHG emissions reduction targets. Freight and heavy good vehicles are also significant contributors to overall transport sector $CO_2$ emissions.[5] Despite improvement in engine efficiency and a slow introduction of biofuels in the last decade, the increasing demand for freight transport (EEA, 2011a) and the dominance of road transport[6] over rail or maritime transport in meeting this demand is mainly responsible for the high emissions profile of road freight transport (Chapter 10).

After road transport, air transport is the second largest contributor of carbon emissions from the transport sector. Domestic aviation and international bunkers were responsible for 12.7 per cent of transport-generated GHG emissions in Europe in 2008 and for 2.6 per cent of $CO_2$ emissions worldwide in 2005 (OECD and ITF, 2010; European Commission, 2011a). Air transport has been growing at a very fast pace too. For example, in

*Source:* The author (2011).

*Figure 11.2 Airplane flying over Oxford, UK and visible linear contrails*

Europe $CO_2$ emissions from international aviation grew by nearly 85 per cent between 1990 and 2004 (EEA, 2008; Annela, 2010).

Aviation $CO_2$ emissions vary between different types of aircraft and therefore across airlines and airports. However, aviation contribution to climate change is not only related to $CO_2$ emission but also includes emissions of nitrogen oxides ($NO_x$), sulphur particulates and aerosols; emission of radiatively significant particles such as soot or water vapour into the upper atmosphere; and increased cloudiness in the form of persistent linear contrails and cirrus clouds (Lee et al., 2009; Annela, 2010; Love et al., 2010).

Aviation's impact on climate change, per unit of GHG emission, is higher than other modes of transport due to the height at which emission occurs and its contribution to the formation of cirrus clouds and contrails (Figure 11.2).[7] This attribute of aviation, together with the forecasted growth of aviation (passenger traffic is expected to double every 15 years, Lee et al., 2009; Airbus, 2011) and the additional expansion of the low cost airline industry (Chapman, 2007), all constitute a major concern over the future increased contribution of the transport sector to $CO_2$ emissions.

Sea or maritime transport is the least significant contributor to GHG emissions and the most efficient per tonne-km. Maritime transport moves more than 70 per cent of the world's cargo (Inderwildi, 2010) but accounts for only 3 per cent of global GHG emissions (European Commission, 2011b). Shipping in rivers and lakes – inland waterways – has a very small share of passenger and freight transport (only 5 per cent in Europe) and therefore its impact on global emissions is minimal – less than 1 per cent of all transport emissions (Uherek et al., 2010). However, growth in global trade is expected to significantly increase maritime transport in the near future. If no reduction policies are put in place, emissions from maritime transport are estimated to grow by 150–250 per cent in 2050 (Love et al., 2010; OECD and ITF, 2010; Lindstad et al., 2011).

Every year 2.2 trillion passengers and 8.6 trillion tonnes of freight are transported by train around the globe (Uherek et al., 2010). Rail-associated GHG emissions vary according to train type, energy source,[8] operating conditions and the service provided (speed, route topography, number of stops and distance between them and so on; Givoni et al., 2009). When electric power is not available, trains usually use fossil fuels, namely, diesel oil, which is responsible for higher impacts in terms of GHG emissions (Uherek et al., 2010).

The transport sector contribution to overall emissions is not just restricted to vehicle operations and fuel combustion. The energy and emissions required during the production of vehicles, provision of infrastructure or fuel production have a significant impact on the environment too. A life cycle assessment or analysis (LCA) is a comprehensive approach that evaluates the environmental impacts from transport by taking into account the operation and non-operation components (such as vehicle manufacturing and disposal; infrastructure construction, operation and maintenance; or petroleum exploration, refining and fuel distribution, among others).

The large requirement of concrete for transport infrastructure contributes significantly to $CO_2$ emissions, produced during cement manufacturing (Chester, 2008).[9] Similarly, the electricity required for vehicle manufacturing, infrastructure materials production and fuel refining and transport generate large amounts of $CO_2$ and sulphur dioxide ($SO_2$). For example, in roadway construction asphalt is responsible for the emission of volatile organic compounds and fugitive dust releases associated with emission of particulate matter $PM_{10}$ (Chester et al., 2010). In 2009 Chester and Horvath (2009) estimated that the total life cycle energy inputs and GHG emissions contribute an additional 63 per cent for road transport, 155 per cent for rail transport and 31 per cent for air systems over vehicle tailpipe operation. At the same time, other studies argue that vehicle production and disposal accounts for 23 per cent of vehicles' emissions life cycle (EEA, 2010) or that that tailpipe emissions in freight transport are underestimated by around 38 per cent for $CO_2$ and $NO_x$ emissions[10] (Facanha and Horvath, 2007). Using this approach at an urban metropolitan level (San Francisco, Los Angeles and New York) it was found that the inclusion of life cycle processes results in an increase of emissions by 20 times the vehicle operation for the different modes in that region (Chester et al., 2010).

**Transport and Air Pollution**

Emissions of GHGs are not the only undesirable impact of the transport sector on the environment. Besides GHG emissions, transport is similarly accountable for the emission of toxic local pollutants, as $NO_x$, carbon monoxide (CO), hydrocarbons (HC), ozone ($O_3$), lead (Pb), $SO_2$ and particulate matter of various sizes ($PM_{10}$ and $PM_5$) (EEA, 2009). Various studies have examined the impacts of air pollution on human health.[11] According to the World Health Organization (WHO) there are more than two million premature deaths per year due to urban outdoor air pollution and indoor pollution from the burning of solid fuels (WHO, 2006). Air pollutants are associated with respiratory and cardiovascular diseases, cancer (related with emissions of benzene) and loss of life expectancy (in case of long exposure to particulate matter), among others. There is agreement that these pollutants represent a serious risk for human health (Uherek et al., 2010).

In various regions of the world, such as Europe (Banister, 2011a) and especially developing countries (Uherek et al., 2010), transport has been associated with the increase in air pollution. Urban air pollution is considered a major health hazard and is estimated to cause 750000 deaths in the world every year (Woodcock et al., 2007).

Land transport has a considerable negative influence on air quality (Figure 11.3). In Europe road transport remains the primary source of $NO_x$ and the second most important source of fine particulates ($PM_{10}$), both associated with high impacts on air quality and human health (EEA, 2009; Uherek et al., 2010). In addition to tailpipe emissions of air pollutants from vehicle operation, air pollution can also increase as a result of congestion, tyre and brake wear, diesel vehicles fitted with oxidation catalysts (which can produce excess $NO_2$ by product) or an increasing number of short trips under cold-start conditions (Woodcock et al., 2007; EEA, 2009).

Aviation contributes to air pollution through the emission of various gases and particles such as $NO_x$, sulphate particles and soot particles. Although fuel combustion in aircraft engines and jet exhausts is efficient and almost smoke free, emissions from aircraft at ground level (considered by the International Civil Aviation Organization (ICAO) to be up to 3000 ft, about 915 m) are increasing with airport movement and associated traffic. A significant quantity of $NO_2$ is emitted at ground level during low power conditions (idle) (Lee et al., 2010), but around some major airports it is surface transport and not aircraft that contributes most to air pollution (Carslaw et al., 2006).

International shipping emits significant quantities of PM, $SO_x$ and $NO_x$, usually more concentrated around ports and coastal areas. Due to fore-

*Source:* The author (2011).

*Figure 11.3* *In great parts of cities around the world cars are the dominant mode of transport. As an example, a main street in Madrid, Spain, dominated by cars*

casted growth in global trade, maritime transport is expected to contribute even more significantly to local air pollution (EEA, 2009).

Rail transport share of emissions is relatively small (1–3 per cent) and can vary significantly depending on the trains' power source (diesel or electricity) or age (UIC and CER, 2006). Electric-power trains could have almost zero impact on local air quality since emissions do not take place along the tracks (Givoni et al., 2009). However, terminal stations with a high proportion of diesel traction will have a significant concentration of $NO_2$ and $PM_{10}$ (UIC and CER, 2006). The age of the train will also influence the quantity of air pollutants emitted; recent locomotives and railcars emit significantly less $NO_x$ and PM compared to older trains.[12]

As noted above, air pollution caused by the different modes of transport varies spatially. Thus, road vehicle emissions have more influence in urban areas with high population densities, with air transport having a greater impact near airports and maritime transport near the coast and around the ports.

**Transport and Noise**

Today, most people live principally in agglomerations and urban areas (Chapter 5), close to transport infrastructure. This means that in many

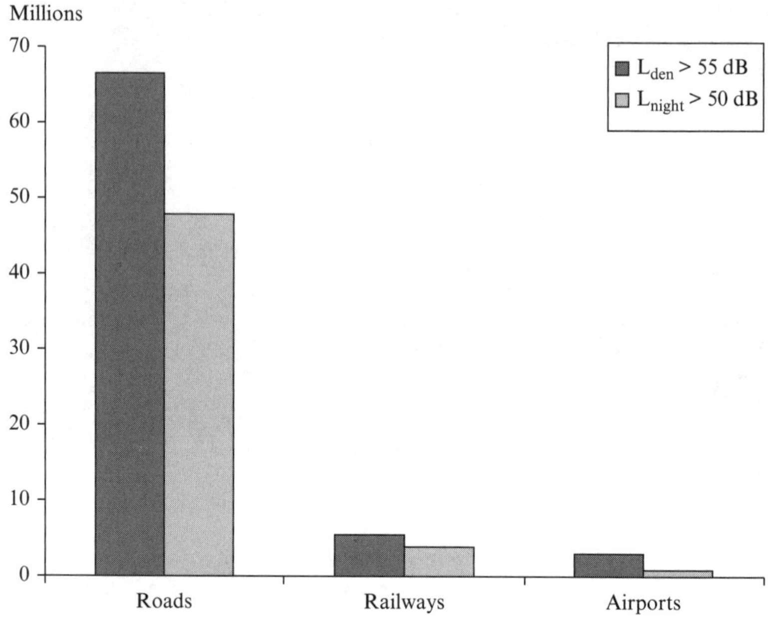


*Source:* EEA (2012).

*Figure 11.4 People affected by transport noise in agglomerations in Europe (EU-27)*

cases citizens are exposed to high levels of noise from transport, which affects their quality of life and can be responsible for long-term health effects (EEA, 2009). Noise pollution in general and especially road and air traffic noise has been associated with disturbance of ecosystems and wildlife along with impacts on health. Effects on humans range from sleep disturbance and elevation of blood pressure to risk of cardiovascular disease or effects on mental health. Some studies have identified that high levels of road traffic noise at night can increase levels of hormones in children and women, or that transport noise was linked with an increase of rates of psychological problems. In children the most common effect of high exposure to noise is cognitive impairment affecting reading comprehension and recognition memory (EEA, 2009).

Noise from road traffic is considered the main source of noise exposure from transport (Figure 11.4). In Europe 55 per cent of the population living in urban areas with more than 250000 inhabitants is exposed to daily road noise levels exceeding 55 Lden.[13] Road traffic noise results from three basic sources: power-train noise (cooling-fan, engine, drive-train and exhaust), tyre-road interaction noise and wind noise. Power-train noise and tyre road interaction vary accordingly to vehicle type, driving style and road surface. Drag noise is mainly important at high speeds. The trend of cars having wider tyres is an example of an action that actually increases rolling noise and resistance (EEA, 2009).

Exposure to railway and airport noise is lower but yet significant where it occurs. Populations living near railway lines can suffer from a substantial exposure to noise. Rail freight is the main contributor to railway noise, and the impact of high speed trains and inner urban railways is also significant (EEA, 2009; European Commission, 2011a).

Aviation also plays a significant role in terms of noise pollution. Recent estimates for Europe have shown that the number of people exposed to air traffic noise has increased, particularly at night. Aircraft's landing and departing, 'holding' around airports and increased use of airports' capacity all contribute to these high noise exposure levels (MPD, 2007; EEA, 2009).

**Transport, Biodiversity and Wildlife**

Transport infrastructure is a predominant feature in many landscapes, and for this reason significant changes to ecological communities and animal populations have been associated with the transport sector. These impacts are quite diverse and can range from mortality, habitat loss and fragmentation to audiovisual disturbance and chemical pollution (Glista et al., 2009; Kociolek et al., 2011).

Networks of roads, in particular, have been associated with habitat loss and fragmentation, as they physically destroy habitats and work as barriers, limiting movement and dispersion of species, from amphibians to large mammals (Lewis et al., 2011; Figure 11.5). Moreover, smaller nature areas can contain less numbers of species and hence reduce biodiversity (Dengler, 2009). Vehicle-caused mortality is also a major source of direct mortality in some animal populations along with pollution and poisoning with road salt, petroleum-based organic compounds or other substances that run off from paved roads (Glista et al., 2009). Furthermore, traffic noise and artificial light can also have negative impacts in birds' communication, habitat quality, migrating and nestling patterns (Kociolek et al., 2011).

Airports are types of transport infrastructure that can impact negatively on wildlife too. As the footprint of an airport on local habitats changes via operations, land management or expansion, the wildlife communities using those habitats are unavoidably affected in their richness and abundance (Blackwell et al., 2009). Bird strikes into aircraft have now

*Source:* The author (2010).

*Figure 11.5 Example of a road working as a barrier and impacting on habitat quality in Patagonia, Argentina and Los Angeles, US*

a higher likelihood of happening due to an increase in air traffic and aircrafts' greater surface area, including bigger engines and associated vacuum (Blackwell et al., 2009). Maritime transport can also negatively affect biodiversity and marine life. Routine shipping operations and especially large-scale accidental oil spills pose a risk to the environment and natural habitats. These types of accidents can also represent economic losses, mainly in coastal areas, where concentration of maritime economic activities such as ports or fishing is higher (Ng and Song, 2010).

## Transport, Water Quality and Soil Erosion

Transport activities and operations are also linked with significant impacts on water quality. Land and air transport modes emit and discard products such as fuel, chemical and other hazardous particulates while in operation. Furthermore, transport-related activities such as de-icing or road maintenance also contribute to the disposal of harmful materials and particulates into the soil and water bodies. These processes can contribute to the contamination of rivers, lakes, wetlands and oceans and change water quality and hydrological conditions. Maritime transport impact on water quality is even more significant and can occur through dredging, waste disposal, ballast waters and oil spills. Dredging can negatively affect the marine environment due to the turbidity caused, the disposal of raised water or decontamination of sediments. The waste generated by ships at sea or at ports (frequently not biodegradable or containing high levels of bacteria) and ballast water (often acquired in one region and discharged in a new marine environment) can be hazardous and disruptive for the natural marine ecosystems. Finally, probably the most serious problem of water pollution can occur when oil cargo vessels are involved in accidents leading to major oil spills (Rodrigue, 2009).

Transport activities can also contribute to soil erosion and soil contamination. The implementation of transport facilities in coastal regions or shipping activities in rivers have been associated with damage of riverbanks or coastlines. Moreover, the implementation of transport infrastructure such as roads, railways or airports usually implicates the removal of large quantities of earth, exposing some areas to erosion or leading to loss of fertile soils. As mentioned above, the discharge or use of toxic materials by the transport industry such as the spill of fuels and oil by motor vehicles on roads or the use of chemicals to preserve railway sleepers are examples of processes that can contribute to soil contamination, mainly next to such infrastructure (Rodrigue, 2009).

**Transport and Land Use**

The need for transport infrastructure requires space and change in land use and this can lead to several impacts on the landscape such as fragmentation, loss of fertile soil or biodiversity (Figure 11.6). There is little data available on the share of land that is actually occupied by the different modes of transport. Yet, some figures show that transport infrastructure covers 1.2 per cent of the total area of the EU (EU 15) or 1.5 to 2.0 per cent of the world's total land surface (Rodrigue, 2009). Furthermore, 93 per cent of land taken by transport infrastructure is dedicated to road transport and only 4 per cent and 1 per cent are dedicate to rail and aviation, respectively (EEA, 2002). The ratio between land used and volume of traffic transported can vary significantly between modes. For example, railways usually require less land taken per transport unit than road

*Source:* The author (2011).

*Figure 11.6 Example of space taken by road infrastructure in Los Angeles, US*

transport – land taken per passenger-km by rail is about 3.5 times lower than for cars (EEA, 2002).

Land requirement for transport infrastructure can be more critical in urban areas, where space is more valuable. Urban transport infrastructure (parking space, roads, petrol stations and so on) is frequently demanding more space. Some studies argue that an exponential increase in the road network and traffic volume[14] is anticipated in the near future in large, densely populated areas and countries such as China and India (Bhattacharya, 2008).

Transport also shapes land use patterns as new transport infrastructure affects demand for travel and development. Investment in transport networks around cities promotes the decentralization of facilities and services, which impacts on suburban land use and increases pressure for developments on suburban areas. Urban sprawl is usually unplanned and with low density, making public transport inefficient and promoting the use of private vehicles (EEA, 2009).

In addition, production of biofuels and their potential expansion can also have impacts on biodiversity and can change land use. Meeting ambitious targets of biofuel usage will require improvements in crop yields and expansion of cropland for biofuel production. A heated debate is taking place related to developments in food demand, agricultural production shortfalls and ethical issues concerning the shift of agricultural output from food crops to transport fuels. Although estimates differ, it is agreed that the conversion of forest and grassland areas to meet biofuel targets can make the GHG balance of biofuels negative (EEA, 2009), or can reduce the benefits that the biofuel programmes are intended to create (EEA, 2010; Findlater and Kandlikar, 2011).

## REDUCING TRANSPORT IMPACTS ON THE ENVIRONMENT

The transport sector, a key element in the global economy and in modern society, is the third largest contributor of anthropogenic GHG emissions and responsible for significant impacts on air and noise pollution, land resources, water and soil quality, wildlife and biodiversity. Moving towards a low carbon mobility that has fewer impacts on the environment and is more sustainable is therefore a priority and a challenge. While Part II of this book is dedicated to this challenge, below some measures that reduce impacts of transport on $CO_2$ emissions, air pollution, noise or biodiversity are discussed.

Transport is one of the few sectors of the economy where emissions are estimated to continue growing and is considered by many to be one of the sectors where it is most difficult and expensive to reduce energy demand and GHG emissions (Brand et al., 2010). Despite the implementation of various mitigation measures, GHG emissions from transport are still growing[15] due to increases in transport demand and mobility of people and goods. Modern lifestyles require travel and promote a desire to move faster and over greater distances – the distance travelled is increasing over time as faster transport replaces slower forms of transport, and this seems to be the main cause for high growth in the volume of GHG emissions (Banister, 2011a; Chapter 1).

Moving towards a more sustainable transport system requires the adoption of a wide range of strategies and measures. Technological measures, alternative fuels and travel behaviour change along with transport policies that include pricing packages, strategic land use and transport planning and green supply chain management practices are all essential to reduce the impacts of transport on the environment (Table 11.2). Examples of policy packages and transport regulation strategies designed to reduce transport demand or to facilitate a modal shift towards less carbon-intensive transport system include, in Europe, the European Transport and Environment Reporting Mechanism (TERM), Biofuels Directive, Energy Taxation Directive, Marco Polo Programmes, Railway Packages,

*Table 11.2 Effects of various environmental policy measures on different environment impacts*

| | Local air pollution (human health) | Long-range air pollution (human health, ecosystem damage | Noise | Emissions of GHG | Fragmentation (biodiversity) | Barrier effects |
|---|---|---|---|---|---|---|
| Exhaust emission abatement and cleaner fuel. | + | + | +/– | +/ | 0 | 0 |
| Quieter vehicles, trains, aircraft and ships | 0 | 0 | + | +/ | 0 | 0 |
| Improving energy efficiency of a mode of transport | + | + | +/– | + | 0 | 0 |
| Shift from individual to public transport | + | + | +/– | + | 0 | +/– |
| Renewable fuels | +/– | +/– | 0 | + | 0 | 0 |
| Improved physical planning | + | +/– | + | +/ | + | + |
| Speed reduction | + | + | + | + | 0 | + |
| Demand management and decreased trafffic growth | + | + | + | + | + | + |

*Note:* + = positive effect; 0 = in principle no effect; +/– = effect uncertain/depends on implementation.

*Source:* EEA (2009).

Euro5 emission standards for cars; or in the US the Corporate Average Fuel Economy (CAFE) and in Japan the Top Runner programme.

## Examples of Specific Measures

Below are some of the measures that can be adopted to reduce transport impacts on the environment are described, considering in turn GHG emissions, noise pollution, air pollution and finally wildlife and biodiversity impacts.

### Reducing GHG emissions from transport sector

The main alternatives to reduce GHG emissions from transport are improving vehicle efficiency (lowering energy consumption) (Chapter 7), developing alternative sources of energy (biofuels, hydrogen, electricity (Chapter 8) or promoting a change in travel behaviour (Chapter 6) and reducing travel demand through planning and policy (Chapters 4 and 5).

The development of new technologies and the introduction of alternative fuels can reduce $CO_2$ emissions from road transport. For example, the potential for GHG emissions reduction when compared with internal combustion engine vehicles is 28 per cent for hybrid-electric vehicles and 67 per cent for electric vehicles (Brady and O'Mahony, 2011). A trend of smaller, lighter and more efficient internal combustion engine vehicles will have positive effects on the environment (Schipper, 2011).

In aviation alternative fuels and fuel reduction through air traffic management and technological measures can play a role in reducing emissions. Air traffic management and operational practices, such as lowering flight speeds, optimizing cruise altitudes, reducing vertical separation minimum (RVSM) or reducing delays in holding patterns on arrival can reduce fuel usage (Lee et al., 2009). Other operational measures, such as increasing aircraft size and adjusting the service frequency to offer similar seating capacity, are also expected to decrease aviation impact on climate change, and also decrease noise pollution, but might result in increased air pollution (Givoni and Rietveld, 2010). There is, however, debate about technological measures and mitigation policies, such as the European Emissions Trading Scheme,[16] but with continued expansion of air transport (Chapman, 2007) they could help address the challenge of mitigating aviation impacts on the environment (Anger and Köhler, 2010; Annela, 2010).

As in road transport and aviation, maritime transport GHG emissions reductions can be achieved if lower speeds are adopted, engines are made more efficient and alternative low carbon fuels are developed. Liquefied gas is an option in terms of alternative fuel for maritime transport and has the advantage of reducing both $NO_x$ and $SO_x$ emissions, as well as $CO_2$ emissions (Love et al., 2010).

In rail the substitution of electric for diesel trains can be considered a major energy source improvement, depending on how the electricity is generated.[17] Other technological improvements that can potentially reduce $CO_2$ emissions from rail transport include the re-engineering of existing fleets, changing the trains' aerodynamics, weight reductions and the development of modern hybrid electric or hydrogen cell drive trains. For example, reducing stabling loads,[18] running shorter off-peak trains or having traffic management systems can all save fuel and therefore reduce GHG emissions (Givoni et al., 2009).

The Emissions Trading Scheme (ETS), Clean Development Mechanisms (CDM) and accounting for the Social Cost of Carbon (SCC) in pricing transport services can all have an important role in reducing GHG emissions.

**Reducing Noise from the Transport Sector**

A variety of mitigation measures can be deployed to reduce noise from transport. To reduce noise at source, an inclusive approach that combines technical improvements to vehicles (such as better silencers, low noise tyres or aerodynamics) along with speed moderation and infrastructure improvements is required (EEA, 2009; Kociolek et al., 2011). Moreover, the introduction of electric vehicles and fuel cell vehicles, without power-train/combustion noise can contribute to noise reduction from road transport.

In addition, land use, transport planning and house standards can play an important role in avoiding the future increase in exposure to transport noise (EEA, 2009). Planning measures that promote the separation of heavy traffic from highly populated areas, barriers or improved sound insulation in houses exposed to high levels of transport are some examples of simple measures that can reduce the adverse environmental impacts of traffic. The development of strategic noise maps and a review of noise standards can also support more effective noise regulation for the transport sector.

Noise reduction operation procedures and operating restrictions, such has specifying preferential routes, concentrating flight path in airspace and airports, setting flying over height or continuous descent approaches will support noise mitigation from aviation (EEA, 2009). In rail transport noise mitigation will be more effective if reduced at source and not by placing noise barriers along railways or soundproofing dwellings (EEA, 2009). Examples of mitigation measures at source include having quieter locomotives, controlling wagon-rolling noise, using low noise brakes or wheel maintenance, among others. Noise-related charges and mechanisms designed to incentivize the use of less noisy vehicles have been rarely implemented. One of the few exceptions includes an aircraft noise-related charging scheme introduced in some airports in France, Germany and the UK.

**Reducing transport impacts on air pollution**

To reduce air pollution, specific measures already exist in some countries and have been implemented in various cities across Europe. The introduction of low emissions zones (LEZs) has been one of the main measures to drive a faster fleet renewal and encourage cleaner technologies. Examples of measures taken by cities such as Utrecht, Prague, Athens, London or Stockholm include the introduction of Park and Ride schemes, setting different parking charges based on emissions, replacement of a city's fleet of cars with cleaner vehicles, implementation of tunnels to filter exhaust fumes away from residential areas or restricted access of vehicles in certain parts of the city depending on their environmental performance. Simultaneously, the increase of engine efficiency and the introduction of three-way catalytic converters in passenger cars along with stricter regulation of emissions from HGVs can reduce exhaust emissions of gases such as $NO_x$, HC, CO and PM (EEA, 2008; Piecyk and McKinnon, 2010).

From a different perspective, the separation of pedestrians from road vehicles can, for example, reduce impacts of air pollution on human health. Levels of exposure to air pollution vary depending on the route taken, traffic density and characteristics of the road; pedestrians are usually exposed to greater values of pollutants compared to those in cars (Briggs et al., 2008).

**Reducing transport impacts on wildlife**

Finally, maintaining the functional connectivity of habitats and the permeability of road system is one of the important measures to mitigate transport impacts on wildlife (Glista et al., 2009). Using empirical data to predict high probability crossing areas along major roads and to define landscape planning strategies to maintain habitat connectivity for animals is an important task for both animal conservation and human safety. Wildlife crossing structures are then essential to maintain this connectivity and can range from exclusion fences, pipe culverts to overpass and underpass systems (Glista et al., 2009). These structures facilitate animal movement across roads and can reduce wildlife mortality from vehicle collisions (Lewis et al., 2011). Vegetation management and restoration along main transport infrastructure can also have a positive impact on the environment.

Overall, planning of new transport infrastructure, such as new roads, railways, ports or airports, should always take into account the environment capacity of the area by identifying sources of pollution, minimizing impacts on visual amenities at the same time as defining natural conservation areas and anticipating disturbance fauna habitats (Blackwell et al., 2009).

## FINAL REMARKS

A low carbon and more sustainable transport system will only be achieved if transport uses less and cleaner energy, uses better more modern infrastructure and reduces its impacts on natural assets like water, land and ecosystems (European Commission, 2011b). A better understanding of the fundamental relationships between the physical environment – topography, geology, geographical location, climate, hydrology, natural vegetation and animal life – and transport activities and infrastructure is crucial (Rodrigue, 2009).

Simultaneously, measures and policies that reduce the need for travel, create shorter journeys and promote the use of more efficient ways of travel will be fundamental. Reducing carbon emissions from the transport sector and 'growing out' of oil dependence will not be possible by relying on a single technological solution. A new concept of mobility, supported by a cluster of new technologies, policy options and more sustainable behaviour should be part of the plan to be adopted.

Ambitious strategies need strong leadership and long-term vision to guide the process. Informed strategies and effective planning will reduce GHG emissions from the transport sector and will simultaneously deliver environmental benefits in other areas such as air pollution, noise exposure, land resources, habitats fragmentation and wildlife.

## NOTES

1. Carbon dioxide is the predominant anthropogenic greenhouse gas emitted. The atmospheric concentration of carbon dioxide has increased from a pre-industrial value of about 280 ppm to 379 ppm3 in 2005, clearly overtaking the natural range from the last 650000 years (180 to 300 ppm) (IPCC, 2007).
2. Global $CO_2$ emissions from the transport sector increased 45 per cent between 1990 and 2007 and are expected to continue to grow by 40 per cent between 2007 and 2030 (OECD and ITF, 2010; EEA, 2011a).

3. Recent figures in Europe have shown that 72 per cent of all passenger-km are done using a car (European Commission, 2011a).
4. In Europe car ownership increased by 22 per cent between just 1996 and 2006, which represents 52 million new cars in ten years (EEA, 2009). In developing countries such as India and China and as a result of their current economic growth, the number of cars is expected to double to two billion in the next two decades (Sperling and Gordon, 2009).
5. Heavy good vehicles (HGVs) and light vans account for approximately 30 per cent of overall transport sector emissions in the UK (DFT, 2011).
6. In 2008 in Europe, the volume of freight transported by road was four times higher than the volume transported by rail (Piecyk and McKinnon, 2010).
7. The IPCC estimates that aviation's total climate impact could be two to four times more than that of its $CO_2$ emissions alone. Furthermore, it is estimated that aviation caused 3.5 per cent of human-induced global warming in 2001 (IPCC, 2007) and future radiative forcing may increase, according to scenarios compatible with the *Special Report on Emissions Scenarios*, by three to four times compared to 2000. Although there is uncertainty about the science of high level emissions, recent estimates indicate that aviation represents a 3.5 per cent share of total anthropogenic forcing in 2005 or a 4.9 per cent share when accounting for contrails formation (water vapour emitted at high levels) (Lee et al., 2009).
8. As an example, Swiss railways' trains are powered with a mix of electric energy: 74 per cent of traction is produced by hydroelectric power and the remainder by nuclear power stations. Both energy sources have an overall low impact in terms of carbon emissions (Tuchschmid, 2011).
9. The cement industry contributes approximately 5 per cent of global man-made $CO_2$ emissions. These emissions are largely the result of the calcination process of limestone, combustion of fuels and power generation. Nearly 900 kg of $CO_2$ are emitted for every 1000 kg of cement produced (Mahasenan et al., 2003).
10. Total life cycle emissions in freight transport have shown that CO and $SO_2$ emissions can be up to seven times higher then tailpipe emissions alone (Facanha and Horvath, 2007).
11. For a full list of references see Uherek et al. (2010).
12. In Europe new trains introduced after 1990 reduced $NO_x$ and PM emission factors by at least 30 per cent and 70 per cent, respectively (Uherek et al., 2010).
13. Lden refers to annual long-term average noise level over 24 hours and defines the European Union (EU) benchmark for excessive noise (EEA, 2009).
14. Estimates have suggested that road area is expected to increase by 27 900 $km^2$ by 2030, with an additional 94 100 $km^2$ due to lanes added to existing roadways just in the US (Theobald, 2010). Moreover, increased transport infrastructure in China is expected to occupy an additional 270 000 $km^2$ between 2004 and 2020 and this will affect three to seven million farmers (Woodcock et al., 2007).
15. For example, in Europe continuous improvements in vehicle and engine efficiency, the development and use of biofuel and other initiatives and policies to promote less carbon-intensive transport modes had for the period between 1990 and 2008 a limited impact on $CO_2$ emissions (EEA, 2011b).
16. The European Union Emissions Trading Scheme (EU ETS) is currently the world's largest emission trading system and the first emission trading scheme that crosses country borders (Anger and Köhler, 2010).
17. According to a life cycle study of various electricity energy sources, coal-fired and natural gas-fired plants emit around 960 and 443 grams of carbon dioxide per kilowatt-hour of electricity produced (gCO2e/kWh); nuclear power plants emit 66 gCO2e/kWh; solar photovoltaic emits 32 gCO2e/kWh and wind farms emit around 10 gCO2e/kWh (Sovacool, 2008).
18. The load/energy for lighting, heating and in general providing power for trains when stationary at depots or stabling sides.

# REFERENCES


Airbus (2011), *Global Market Forecast*, Airbus, available at http://www.airbus.com/company/market/forecast/ (accessed 22 June 2012).

Anger, A. and J. Köhler (2010), 'Including aviation emissions in the EU ETS: much ado about nothing? A review', *Transport Policy*, **17**(1), 38–46.

Annela, A. (2010), 'Including aviation in the European emissions trading scheme: impacts on the industry, $CO_2$ emissions and macroeconomic activity in the EU', *Journal of Air Transport Management*, **16**(2), 100–5.

Banister, D. (2011a), 'Cities, mobility and climate change', *Journal of Transport Geography*, **19**(6), 1538–46, available at http://www.sciencedirect.com/science/article/pii/S0966692311001244 (accessed 21 October 2011).

Banister, D. (2011b), 'Transport and environment', available at http://www.annualreviews.org/doi/pdf/10.1146/annurev-environ-032310-112100.

Bhattacharya, A. (2008), 'Linking South East Asia: more connectivity, better ties', Institute of Peace and Conflict Studies, available at http://www.ipcs.org/pdf_file/issue/935268060IPCS-SpecialReport50.pdf (accessed 23 May 2012).

Blackwell, B.F., T.L. DeVault, E. Fernández-Juricic and R.A. Dolbeer (2009), 'Wildlife collisions with aircraft: a missing component of land-use planning for airports', *Landscape and Urban Planning*, **93**(1), 1–9.

Brady, J. and M. O'Mahony (2011), 'Travel to work in Dublin: the potential impacts of electric vehicles on climate change and urban air quality', *Transportation Research Part D: Transport and Environment*, **16**(2), 188–93.

Brand, C., M. Tran and J. Anable (2010), 'The UK transport carbon model: an integrated life cycle approach to explore low carbon futures', *Energy Policy*, **41**, February, 107–24, available at http://www.sciencedirect.com/science/article/pii/S0301421510006348 (accessed 31 October 2011).

Briggs, D.J., K. de Hoogh, C. Morris and J. Gulliver (2008), 'Effects of travel mode on exposures to particulate air pollution', *Environment International*, **34**(1), 12–22.

Carslaw, D.C., S.D. Beeversb, K. Ropkins and M.C. Bell (2006), 'Detecting and quantifying aircraft and other on-airport contributions to ambient nitrogen oxides in the vicinity of a large international airport', *Atmospheric Environment*, **40**(28), 5424–34.

Chapman, L. (2007), 'Transport and climate change: a review', *Journal of Transport Geography*, **15**(5), 354–67.

Chester, M.V. (2008), 'Life-cycle environmental inventory of passenger transportation in the United States', available at http://escholarship.org/uc/item/7n29n303#page-141 (accessed 13 July 2012).

Chester, M.V. and A. Horvath (2009), 'Environmental assessment of passenger transportation should include infrastructure and supply chains', *Environmental Research Letters*, **4**(2), doi 024008.

Chester, M.V., A. Horvath and S. Madanat (2010), 'Comparison of life-cycle energy and emissions footprints of passenger transportation in metropolitan regions', *Atmospheric Environment*, **44**(8), 1071–9.

Dengler, J. (2009), 'Which function describes the species-area relationship best? A review and empirical evaluation', *Journal of Biogeography*, **36**(4), 728–44.

DFT (2011), *Transport Statistics Great Britain 2011*, Department for Transport,

available at http://assets.dft.gov.uk/statistics/releases/transport-statistics-great-britain-2011/tsgb-2011-summaries.pdf.
EEA (2002), *Land Take by Transport Infrastructure – TERM 1 European Environment Agency*, available at http://www.eea.europa.eu/data-and-maps/indicators/land-take-by-transport-infrastructure (accessed 16 February 2012).
EEA (2008), *Climate for a Transport Change. TERM 2007: Indicators Tracking Transport and Environment in the European Union – EEA*, available at http://www.eea.europa.eu/publications/eea_report_2008_1 (accessed 20 May 2011).
EEA (2009), *Transport at a Crossroads. TERM 2008: Indicators Tracking Transport and Environment in the European Union*, Copenhagen: European Environment Agency, available at http://www.eea.europa.eu/publications/transport-at-a-crossroads (accessed 21 October 2011).
EEA (2010), *Towards a Resource-efficient Transport System. TERM 2009: Indicators Tracking Transport and Environment in the European Union*, Copenhagen: European Environment Agency, available at http://www.eea.europa.eu/publications/towards-a-resource-efficient-transport-system (accessed 19 February 2012).
EEA (2011a), *Greenhouse Gas Emissions in Europe: A Retrospective Trend Analysis for the Period 1990–2008 – EEA*, available at http://www.eea.europa.eu/publications/ghg-retrospective-trend-analysis-1990-2008 (accessed 31 October 2011).
EEA (2011b), *Greenhouse Gas Emissions in the EU-27 by Sector in 2008, and Changes between 1990 and 2008 – Maps and Graphs – EEA*, available at http://www.eea.europa.eu/data-and-maps/figures/greenhouse-gas-emissions-in-the (accessed 20 May 2011).
EEA (2012), *The Contribution of Transport to Air Quality – TERM 2012: Transport Indicators Tracking Progress Towards Environmental Targets in Europe*. Copenhagen: European Environment Agency, available at http://www.eea.europa.eu/publications/transport-and-air-quality-term-2012.
European Commission (2011a), *Statistical Pocket Book 2011 EU Energy and Transport in Figures*, European Commission, available at http://ec.europa.eu/transport/publications/statistics/pocketbook-2011_en.htm (accessed 14 February 2012).
European Commission (2011b), *White Paper Roadmap to a Single Transport Area – Towards a Competitve and Resource Efficent Transport System*, European Commission, available at http://eur-lex.europa.eu/LexUriServ/LexUriServ.do?uri=CELEX:52011DC0144:EN:NOT (accessed 14 May 2011).
Facanha, C. and A. Horvath (2007), 'Evaluation of life-cycle air emission factors of freight transportation', *Environmental Science and Technology*, **41**(20), 7138–44.
Findlater, K.M. and M. Kandlikar (2011), 'Land use and second-generation biofuel feedstocks: the unconsidered impacts of Jatropha biodiesel in Rajasthan, India', *Energy Policy*, **39**(6), 3404–13.
Forster, P. and V. Ramaswamy (2007), 'Changes in atmospheric constituents and in radiative forcing', in S. Solomon, D. Qin, M. Manning et al. (eds), *Climate Change 2007: The Physical Science Basis. Contribution of Working Group I to the Fourth Assessment Report of the Intergovernmental Panel on Climate Change*, Cambridge: Cambridge University Press.
Fuglestvedt, J., T. Berntsen, G. Myhre, K. Rypdal and R. Bieldvedt Skeie (2008), 'Climate forcing from the transport sectors', *Proceedings of the National Academy of Sciences of the USA*, **105**(2), 454–8.

Givoni, M., C. Brand and P. Watkiss (2009), 'Are railways "climate friendly"?', *Built Environment*, **35**(1), 70–86.
Givoni, M. and P. Rietveld (2010), 'The environmental implications of airlines' choice of aircraft size', *Journal of Air Transport Management*, **16**(3), 159–67.
Glista, D.J., T.L. DeVault and J.A. DeWoody (2009), 'A review of mitigation measures for reducing wildlife mortality on roadways', *Landscape and Urban Planning*, **91**(1), 1–7.
IEA (2009), *Transport, Energy and $CO_2$: Moving Toward Sustainability*, IEA, available at http://www.iea.org/publications/free_new_Desc.asp?PUBS_ID=2133 (accessed 16 May 2012).
IEA (2010), *$CO_2$ Emissions from Fuel Combustion*, Paris: OECD and IEA, International Environment Agency, available at http://www.iea.org/co2highlights/co2highlights.pdf (accessed 17 May 2012).
Inderwildi, O. (2010), *Future of Mobility Roadmap: Ways to Reduce Emissions while Keeping Mobile*, Oxford: Smith School of Enterprise and the Environment.
IPCC (2007), *A Report of Working Group I of the Intergovernmaentla Panel on Climate Change – Summary for Policemakers*, Cambridge and New York: Cambridge University Press.
Kociolek, A.V., A.P. Clevenger, C.C. St Clair et al. (2011), 'Effects of road networks on bird populations', *Conservation Biology*, **25**(2), 241–9.
Lee, D.S., D.W. Fahey, P.M. Forster et al. (2009), 'Aviation and global climate change in the 21st century', *Atmospheric Environment*, **43**(22–23), 3520–37.
Lee, D.S. et al. (2010), 'Transport impacts on atmosphere and climate: aviation', *Atmospheric Environment*, **44**(37), 4678–734.
Lewis, J.S., J.L. Rachlow, J.S. Horne et al. (2011), 'Identifying habitat characteristics to predict highway crossing areas for black bears within a human-modified landscape', *Landscape and Urban Planning*, **101**(2), 99–107.
Lindstad, H., B.E. Asbjørnslett and A.H. Strømman (2011), 'Reductions in greenhouse gas emissions and cost by shipping at lower speeds', *Energy Policy*, **39**(6), 3456–64.
Love, G., A. Soares and H. Püempel (2010), 'Climate change, climate variability and transportation', *Procedia Environmental Sciences*, **1**, 130–45.
Mahasenan, N., S. Smith and K. Humphreys (2003), 'The cement industry and global climate change: current and potential future cement industry $CO_2$ emissions', in J. Gale and Y. Kaya (eds), *Proceedings of the 6th International Conference on Greenhouse Gas Control Technologies*, 1–4 October 2002, Kyoto: Elsevier, pp. 995–1000, available at http://www.sciencedirect.com/science/article/pii/B9780080442761501574 (accessed 13 July 2012).
MPD (2007), *Study of Aircraft Noise Exposure at and Around Community Airports: Evaluation of the Effect of Measures to Reduce Noise*, London: European Commission.
Ng, A.K.Y. and S. Song (2010), 'The environmental impacts of pollutants generated by routine shipping operations on ports', *Ocean and Coastal Management*, **53**(5–6), 301–11.
OECD and ITF (2010), *Reducing Transport Greehouse Gas Emissions Trends & Data 2010*, OECD and ITF, available at http://www.internationaltransportforum.org/Pub/pdf/10GHGTrends.pdf.
Piecyk, M.I. and A.C. McKinnon (2010), 'Forecasting the carbon footprint of road freight transport in 2020', *International Journal of Production Economics*, **128**(1), 31–42.

Rodrigue, J.-P. (2009), *The Geography of Transport Systems*, 2nd edn, London: Routledge.

Schipper, L. (2011), 'Automobile use, fuel economy and $CO_2$ emissions in industrialized countries: encouraging trends through 2008?', *Transport Policy*, **18**(2), 358–72.

Sovacool, B.K. (2008), 'Valuing the greenhouse gas emissions from nuclear power: a critical survey', *Energy Policy*, **36**(8), 2950–63.

Sperling, D. and D. Gordon (2009), *Two Billion Cars: Driving Toward Sustainability*, New York: Oxford University Press.

Stern, N. (2006), *Stern Review Final Report*, available at http://webarchive.nationalarchives.gov.uk/+/http://www.hm-treasury.gov.uk/stern_review_report.htm (accessed 16 May 2011).

Theobald, D.M. (2010), 'Estimating natural landscape changes from 1992 to 2030 in the conterminous US', *Landscape Ecology*, **25**, 999–1011.

Tuchschmid, M. (2011), *SBB Ecocalculator – Background Report – Version 1.1*, Zurich: SBB.

Uherek, E., T. Halenka, J. Borken-Kleefeld et al. (2010), 'Transport impacts on atmosphere and climate: land transport', *Atmospheric Environment*, **44**(37), 4772–816.

UIC and CER (2006), *Rail Diesel Emissions – Facts and Challenges*, Community of European Railway and Infrastructure Companies.

Unruh, G.C. (2000), 'Understanding carbon lock-in', *Energy Policy*, **28**(12), 817–30.

WHO (2006), *Air Quality Guidelines for Particulate Matter, Ozone, Nitrogen Dioxide and Sulfur Dioxide – Global Update 2005*, Geneva: World Health Organization, available at http://www.who.int/phe/health_topics/outdoorair_aqg/en/ (accessed 27 September 2012).

Woodcock, J., D. Banister, P. Edwards, A.M. Prentice and I. Roberts (2007), 'Energy and transport', *Lancet*, **370**(9592), 1078–88.

# 12. Mobility as a complex system: key elements and interactions

**James Macmillen**

## INTRODUCTION

The preceding ten chapters have each offered a series of compelling insights into various facets of contemporary mobility. Quite clearly, these various aspects of transport and mobility warrant detailed investigation and the production of specific knowledge surrounding their nature and evolution. Such are their details and uncertainties, academic research quite rightly attends to their individual issues and circumstances. For example, entire academic careers are focused on the nuances and complexities of transport governance within particular countries or at particular jurisdictional scales, and similar professional effort and expertise is directed at understanding the dynamics of biofuel markets and processes of urbanization in the Global South. This is quite necessary, for without the detailed expert knowledge produced by such specialist endeavours we would not have the requisite information required to be able to support societal understanding of transport and mobility more broadly, and importantly, we would not be able to pursue informed, evidence-based policy-making which is critical in shifting contemporary mobility systems onto a more sustainable path. However, while the production and dissemination of such specialist knowledge is absolutely essential, there is also an attendant requirement for broader, 'meta-level' analysis in the mobility and transport domain that duly acknowledges the complex interactions between these various elements and develops a comprehensive, systemic understanding of contemporary mobility practices, their roots in economic and societal practices, and their overall implications for economic, societal and environmental priorities.

To this end, this chapter attempts three things. First, drawing on the contents of the preceding ten chapters, it aims to summarize their principle arguments and conclusions. Most simply, these relate to the need to conceptualize contemporary mobility in more than utilitarian terms; to appreciate its expanding nature and the material and political consequences of transport growth; to acknowledge the major environmental implications of this continued growth; to critically understand novel developments in transport and related technologies; to situate contemporary mobility practices in the broader contexts of globalized economic production and consumption regimes; to grasp the interplay between transport and (changing) urban structures; to appreciate the importance of accessibility and how this varies across sociodemographic groups; to acknowledge the fragmented and messy nature of mobility governance, to appreciate the intricacies and irrationalities of behaviour change; and to understand the dominant processes of transport finance and investment. Second, drawing on secondary social science literature on complexity theory and mobility – and seeking to preface Part II of the book, in particular Chapter 14 on sociotechnical transitions – the chapter stresses the importance of foregrounding the interrelationships between these various facets, aspects or 'elements' of mobility and discusses a few of these linkages in further detail. Third, the chapter critically reflects on particularly dominant framings of these interrelationships that are apparent in current policy discourses and considers the extent to which these interpretations of contemporary mobility practices and systems are conceptually sufficient for supporting a genuine transition to sustainable mobility.

## THEMATIC INSIGHTS

The preceding chapters offer numerous insights into the dynamics of contemporary mobility, the key challenges involved in sufficiently understanding and governing transport practices, and the manner in which they may evolve in the future across geographical contexts. Our ultimate point of departure, as detailed in Chapter 1, concerns the fact that 'mobility' pertains to far more than simply 'transport'. Undoubtedly, mobility partly refers to the empirical 'realities' of movement per se, yet – as the literature on 'mobilities' emphasizes – contemporary travel practices also relate profoundly to issues of representation, social identity and embodied experience (Cresswell, 2006; Chapter 2). Critically, appreciating this more than utilitarian character of mobility is essential if we are to gain a genuine understanding of contemporary transport practices and, on this basis, develop a sustained capacity for effective transport governance. The case of private automobility and the emergence of the motorcar as an iconic cultural commodity is particularly illustrative here. In their purest sense, cars clearly facilitate movement from A to B, often in a relatively flexible and comfortable manner. However, in a multitude of ways, they are also bound up in complex social connotations relating to wealth, gender and sexuality (Miller, 2001; Urry, 2007).

Highlighting the sociocultural complexity of mobility's determinants, however, should in no way serve to obscure or dispute its material and political consequences. Globally, travel is becoming more frequent and trips are becoming longer (Chapter 1). However, while in the Global North there is some evidence that automobility may be 'peaking' (Metz, 2010; Millard-Ball and Schipper, 2011), rising economic growth and car ownership in the emerging economies of the Global South indicate that private vehicle travel is likely to increase substantially in the coming decades, reinforced by the emergence of very low-cost cars, such as the Tata Nano (see O'Connor, 2008).

As discussed in Chapters 1 and 11, this growth in transport activity has a multitude of adverse environmental effects, at both local and global scales. In addition to generating greenhouse gas (GHG) emissions from fuel combustion, contemporary transport practices can cause significant air pollution, noise, ecosystem and biodiversity damage, and water pollution (for an overview, see Banister, 2005 and Chapter 11). In the context of low carbon mobility, however, it is the emissions of GHGs that are the focus for attention. Strikingly, the global transport sector is 94 per cent dependent on oil and moving away from this dependence lies at the heart of any genuine transition to sustainable mobility. The powerful environmental arguments for such a shift in energy supply are complemented through pragmatic concerns regarding 'peak oil' and dwindling reserves of readily accessible crude oil. As outlined in Chapter 8, electricity derived from renewable sources and/or electricity decarbonized at the point of generation represents a much cleaner and more secure form of energy supply than oil. However, significant investment would be required to ensure that global electricity grids are of a sufficient quality to support the massive requirements of the transport sector. Biofuels, natural gas and hydrogen are also potentially viable transport fuels. However, biofuels suffer from a range of environmental problems, there are financial, infrastructural and environmental uncertainties associated with natural gas extraction, and hydrogen remains expensive to produce, store and distribute.

There is also a diverse array of transport technologies that may support efforts to bring about low carbon mobility across various transport modes, as outlined in Chapter 7. Enhancements to vehicle technologies have resulted in improved engine efficiency gains, mass reduction and friction minimization. New operational systems are also emerging, such as regenerative braking, as well as completely novel developments in alternative propulsion mechanisms and aircraft design. Developments in telecommunications are rapidly increasing too, with telepresence technologies increasingly framed as a substitute for travel in their own right. However, all of these technological developments need to be understood as part of a wider system in which their ostensible benefits are appraised in relation to the broader systemic effects of their introduction. Rebound effects are ever-present; improved fuel economy may encourage the travelling of greater distances, and the networking facilitated by video conferencing may generate additional trips, as social ties are maintained and new travel opportunities arise.

While much debate in the transport literature is centred on technological developments, changes in the broader contextual landscape of contemporary transport are vital to consider. Not least, the economic geography of globalization has resulted in substantial increases in international and domestic freight transport. As argued in Chapter 10, the growth of the post-Fordist 'just-in-time' economy has led to the emergence of complex, time-critical and intercontinental supply chains, with resources and commodities travelling huge distances between sites of production and consumption. These developments, and their rapid acceleration, pose major obstacles to a transition to sustainable mobility. The vast majority of freight tonnage is carried by maritime transport, with air transport covering the logistics market for high-value, low-bulk goods. On land road transport is the major form of distribution, although policy efforts are often aimed at shifting a substantial proportion of land-based freight onto the rail network and inland waterways, particularly in Europe. Telecommunications developments and e-commerce are rapidly shaping the nature of logistics, on both the supply side and demand side.

Similarly contextual, Chapter 5 notes how processes of urbanization, particularly in the Global South, are proceeding at a rapid rate. While cities are central to fostering collaboration, information flow and economic activity, associated problems of urban transport are becoming magnified, with pronounced levels of congestion, air pollution and associated environmental implications. Although cities across the world differ enormously, both in form and function, the interrelationships between urban form and transport are abundantly clear; urban form provides a context for transport, but in turn transport comes to shape urban form – most evidently through private car ownership and suburban sprawl. Urban planning, while often subservient to other policy domains, can nevertheless improve the density, diversity and design of cities, thereby facilitating more sustainable modes of urban travel.

In furthering debates on low carbon mobility, much of which is explicitly or implicitly concerned with reducing transport demand per se, it is important not to lose sight of why mobility matters. As explained in Chapter 3, simply put, transport, alongside land use, is a key enabler of accessibility – the extent to which individuals are able to access activities and destinations. In order to understand and improve accessibility through policy interventions, it is important to recognize that individuals' temporal, financial and physical capability constraints will differ significantly across the population. As the chapter notes, equity considerations are clearly central to debates on accessibility; transport-related social exclusion disproportionately affects the young, the elderly, ethnic minorities and those with a disability.

A balance must be struck, therefore, which enables accessibility and mitigates environmental and other harms, and it is this balancing act which represents the guiding rationale for government intervention in the transport system through associated policies and instruments. As Chapter 4 makes clear, however, transport governance is a complex, messy and fragmented affair, characterized by the presence of numerous 'advocacy coalitions' with very different ideas about the purpose and importance of transport. As a result, transport policy at all jurisdictional scales is intensely political in nature. It involves far more than technical planning and often exhibits a lack of integration with closely related policy domains. Moreover, transport governance suffers from a degree of narrowness with regard to its understanding of transport practices. It is deeply entrenched in a utilitarian interpretation of mobility and is rarely capable of taking a systemic, holistic perspective on the broader impacts of transport interventions over time. Policy measures that are typically favoured are those relating to efficiency improvements, which rarely encounter the political opposition that arises with attempts to manage travel demand.

Critically, when dealing with questions of intervention, regulation and governance, understanding processes of behaviour change is central. As argued in Chapter 6, traditional, neoclassical approaches to the study of travel behaviour, which assume that individuals' travel behaviour is invariably 'rational', exhibiting context-independent, temporally stable preferences, are insufficient for fully understanding travel and instigating effective behaviour change (see also Banister et al., 2011). Indeed, it is abundantly clear that habits, perceptions, social identities, aspirations and norms are major determinants of mobility practices. For example, considering the uptake of (hybrid) electric vehicles, it is evident that consumers are motivated only in part by reduced running costs; issues of individualism, shared norms and technological valorization are also prominent (Sovacool and Hirsh, 2009).

Finally, again heavily linked with issues of governance, Chapter 9 highlights the centrality of finance and investment in the context of a transition to low carbon mobility. Investment in transport infrastructure is seen as a core means of stimulating economic growth, although the interrelationship between these two variables is complex and yet to be fully understood. Widely framed as a public good, transport infrastructure is generally funded through public accounts. Increasingly, user charging and public-private partnerships are seen as viable sources of finance – particularly in the case of capital-intensive projects. At present, the transport sector is poorly represented in global climate change finance debates and green growth initiatives. This underrepresentation illustrates the profound uncertainties regarding returns on investment from nascent sustainable transport schemes. Moreover, even where green growth schemes appear successful, issues of rebound effects and unintended consequences remain.

## INTER-THEME LINKAGES

To a degree, then, we can understand the ten themes discussed above to represent the core elements underpinning contemporary mobility systems and transport practices. As stressed in the introduction to this chapter, it is hence vital to recognize the value of conducting detailed empirical research and generating specialist knowledge in relation to each of these elements; understanding their nature and respective dynamics is a precursor to effective intervention and potential transformation. However, it is of critical importance to consider the manner in which these various elements interact with one another in a systemic sense. Most significantly, this is because systems of contemporary mobility represent 'more than the sum of their parts' (cf. Jervis, 1997). In short, they are what a number of theorists term 'complex' or 'open' systems, which – although exhibiting tangible degrees of stability and permanence – are nevertheless inherently unstable and prone to unpredictable, yet transformational, change (Geyer and Rihani, 2010).

Echoing 'post-Newtonian' trends in the natural sciences, and reflecting the broader rise of post-structuralist thinking, recent years have seen a proliferation of academic scholarship in the social sciences seeking to examine a variety of phenomena through a 'complexity' lens. Critically, given their sociocultural foundations and humans' self-awareness, these phenomena can be interpreted as 'conscious complex systems' (Geyer and Rihani, 2010). The literature on such systems is burgeoning, and this chapter is evidently not the place for a comprehensive overview of the field. However, it is worth briefly stating two key principles of significance for understanding contemporary transport and its governance. First, complex systems exhibit degrees of 'partial order' where, despite the propensity for chaos within the system, sufficient order can maintain a particular equilibrium. We can see this, for example, in the present systems of private car manufacturing, ownership and usage, which although having greatly increased in scope throughout the twentieth century, are still essentially based upon the internal combustion technologies and relatively simple infrastructure (contrast this with modern telecommunications, for example). Indeed, drawing heavily on complexity theory, Urry (2004) offers a holistic interpretation of the modern car system using the concept of 'automobility'. Most notably, his work draws attention to the array of feedback processes operating through the system such that particular aspects come to dominate. Central to his argument is the notion of *autopoiesis*, Greek for 'self-creation', whereby the spatially and temporally fragmented nature of modern society – itself heavily determined by the rise of private motorized travel and suburban development – creates the conditions for the perpetuity of that same system, thereby reinforcing or 'locking-in' already dominant patterns and practices. Second, complex systems are characterized by the presence of non-linear relationships between their components, and the associated property of 'emergence'. The condition of non-linearity refers to situations whereby relatively small changes in one area of a system can, by virtue of feedback, lead to substantial changes in the system as a whole. Somewhat related, emergence refers to the materialization of a genuinely new phenomenon as a result of the interaction of existing phenomena. It is a widely understood concept in social science, with Durkheim (1938, p. 157) noting how: 'whenever certain elements combine and thereby produce, by fact of their combination, new phenomena, it is plain that these new phenomena reside not in the original elements but in the totality formed by their union'.

Taken together, these characteristics of complex systems pose incredible challenges for mobility governance and the formulation of effective, efficient transport policy interventions. Of course, transport has always been 'complex'; the transition from horse transport to car transport in the early twentieth century, for example, was a complicated process of transformational change involving a wide array of actors, institutions and sociocultural practices and norms (Geels, 2005). In the past, however, transport governance has tended to restrict itself to responding to the output of complex systems – particularly apparent, for example, in the era of 'predict and provide' where the central task for government was considered to be the provision of ever more transport infrastructure to accommodate growing demand (see Goodwin et al., 1991; Headicar, 2009). However, in the present era, when – in rhetoric at least – priorities have widened to encompass a greater array of economic, societal and environmental objectives, transport policy finds itself increasingly concerned with the dynamics of complex systems, as evident in struggles to engender behaviour change towards 'sustainable' or 'smarter' travel choices.

In order to develop a more systemic appreciation of contemporary mobility, its major challenges and associated policy options, it is fruitful to explore the multiple interactions and linkages that exist between its various components. Such linkages can take many different forms, with their interpretation further depending upon the particular conceptual lens through which the mobility system is considered. The remainder of this section highlights some of the major pair-wise linkages considered to exist between the ten themes discussed in the preceding chapters.[1]

When issues of culture and societal norms are authentically integrated into discussions of transport and mobility, this tends to be focused primarily in relation to passenger transport (for example, Adey et al., 2007; Merriman, 2007). Hence, exploring the linkages between mobility cultures and supply chains and logistics is highly worthwhile and potentially insightful. In the main, cultures of rationality and cost-minimization dominate global supply chain management. In line with the maxim 'time is money', the valorization of speed and efficiency is incredibly pronounced, often reinforced through the prevalence of, and reliance upon, information and communications technologies (ICT). However, alongside these obvious linkages, supply chains and mobility cultures are connected in more subtle ways. There are interesting questions relating to the role of habits and norms and the extent to which particular logistics paradigms – such as 'just-in-time'– are inevitable, or whether they represent a particular business culture that is inherently unstable and replaceable. Additionally, contemporary supply chains are often extremely long-distance and disjointed, directly and indirectly snaking across cultural contexts that differ significantly by geography and industry, highlighting the heterogeneity of the global economy and the scope therein for emergent change.

The most striking linkage between mobility cultures and governance and policy is that the nature of the former is rarely, if ever, appreciated by the latter. Indeed, as argued in Chapter 2 and reiterated above, contemporary transport policy is locked-in to an overly utilitarian interpretation of mobility that it appears unable to transcend. Aside from the fact that policy is hence unable to grasp and reflect the differing requirements and hopes for mobility that exist throughout and across societies, this linkage between the two themes has a number of critical implications. Not least, it leads to – or at least fails to challenge – the predilection for in-cash, economic policy instruments and the focus upon technologies as change agents. Somewhat paradoxically, these two themes are also linked in two normative senses; the utilitarian interpretation of transport as a derived demand views mobility as a cost to be minimized, while broader political narratives tacitly frame mobility as inherently desirable.

There are numerous linkages too between transport technology and finance and investment. Most obviously, the low carbon mobility technology market is relatively nascent, with a fast-moving investment climate involving a wide array of emerging industry, financial and policy stakeholders. Uncertainty is thus highly prevalent, with particular risks emerging in relation to expected technology life-spans, consumer preferences and resultant market share. More broadly, linkages between these two themes also pertain to the need for a realistic appreciation of the economic power presently invested in incumbent, oil-based transport technologies, and the significant amounts of associated sunk capital tied up in existing infrastructure. Hence, future patterns of investment in transport technology will be hugely shaped, both directly and indirectly, by established interests as well as through more novel institutional approaches. Somewhat related are the linkages between the themes of energy and power mix and finance and investment, as the use of novel energy sources, as replacements for fossil fuels, depends heavily on adequately financed technological developments.

A number of core linkages between transport and the environment and accessibility and equity emerge in relation to the implications of present mobility practices for inter- and intra-generational equity. At the most spectacular level, anthropogenic climate change, exacerbated through GHG emissions from transport, poses a major threat to the wellbeing of future generations, with this threat being unevenly distributed across the globe. Levels of accessibility in an era of regular flooding and extreme weather events, for example, are likely to be highly unequal as different social groups exhibit markedly different adaptive capacities. More immediately, however, the distribution of non-GHG environmental consequences is already raising equity concerns. In major cities, for example, it is the relatively wealthy who emit the majority of transport-related air pollutants but the relatively poor, by virtue of urban social geography, whose health is most adversely affected. Additionally, interesting linkages emerge through transport-related climate change mitigation and adaption efforts. For example, pronounced issues of fairness and justice will impact upon the political feasibility of global wealth redistribution. Moreover, macro-level GHG mitigation efforts – formulated with the best of intentions – may paradoxically serve to heighten social exclusion and relative poverty, particularly as a result of regressive measures aimed at curbing aviation and private car use.

Throughout history cities have always been centres of interaction, innovation and communication, and the linkages between urbanization and behaviour change are thus both interesting and systemically significant. Broadly speaking, the linkages between these two themes are bidirectional. First, behaviour change – manifested in changing mobility preferences and practices – can shape both the future form and function of cities, albeit with differing effects across the globe. In this regard, consumer responses to ever-increasing oil prices and the uptake of new technologies may usher in completely novel forms of urbanity, or herald a return to more 'traditional' forms (for example, neighbourhood-scale, mixed land use). Second, the form and function of cities directly and indirectly shapes our travel behaviour in a multitude of ways. Importantly, in order to strengthen this linkage and improve the viability of urban planning and design as legitimate policy instruments, efforts may be needed to incorporate attitudinal dimensions. Additionally, rapid urbanization in the developing world may lead to greater and more urgent urban planning – offering a window of opportunity to 'design in' sustainable forms of mobility at the urban scale.

The most apparent linkage between behaviour change and governance and policy concerns the capacity of the former to facilitate the latter in instances where it may be appropriate for ensuring a viable transition to low carbon mobility. In this regard, and as noted above, governance and policy must come to appreciate the sociocultural significance and character of mobility in order to formulate and implement policy interventions that are effective, efficient and publicly legitimate. Measures centred upon pricing and technology uptake, while important, are not a panacea; in order to bring about behaviour change, there is a pressing need to explore alternative instruments that consider who may be receptive to which messages, with a direct concern for citizens' attitudes and tacit aspirations (Banister et al., 2011). Furthermore, it may be necessary to extend the sphere of governance into areas of the economy where it is presently absent (for example, regulation of car advertising). More positively, additional linkages relate to the agency of behavioural change and the capacity of emergent social movements to shape the foci and priorities of governance – as has been the case, for example, following the resurgence of cycling in London over recent years.

The linkages between transport and the environment and governance and policy are crucially important in the context of a transition to low carbon mobility. At the highest level, there is clearly a pressing urgency for viable global agreements concerning the mitigation of future GHG emissions and the strengthening of global resilience to climate change impacts. As difficult as it may be, such agreements must strive to encompass the international aviation and maritime sectors. The global economy must similarly come under greater scrutiny, with a regulatory focus upon sites of consumption as well as sites of production. Linkages also point to the need for a shift from cost-driven to carbon-driven models of economic activity, and increased support for the Clean Development Mechanism and associated trading schemes. Linkages also emerge at smaller scales, with increasing roles for localism and regionalism, particularly in relation to the resilience of transport networks and the adaptive capacities of current systems.

## DOMINANT FRAMINGS

Thus far, this chapter has summarized the salient points arising from Chapters 2–11 in the book relating to the respective key elements of contemporary mobility systems, and – drawing on the output of an expert workshop – has highlighted the complexity of modern transport systems, illustrated through a discussion of key, inter-theme linkages. Finally, in this third section, the chapter once again draws on the content of the 2011 expert workshop held in Oxford to critically discuss the manner in which these various themes and linkages appear to be framed in contemporary policy-making, and to a slightly lesser degree, in academic research on transport. Workshop participants were invited to discuss what they understood to be the prevailing interpretative mindset present in contemporary policy and academic circles, and to subsequently consider the degree to which the various ten themes or 'elements' of present mobility systems are (a) sufficiently appreciated and represented in prevailing discourses and (b) sufficiently acknowledged in particular policies and studies. The representation of the group's collective argument (Figure 12.1) is inevitably partial; it is neither fully 'positive' – insofar as is does not pretend to offer a comprehensive description of dominant framings, themes and linkages – nor is it at all normative – as is does not aim to encapsulate what 'ought' to be. This said, however, it does represent an interesting attempt to consider the prevailing meta-level policy framings of the transport system, albeit one predominantly based around participants' experiences and professional expertise on transport governance in the Global North.

At the bottom of the figure, processes of urbanization and supply chain dynamics were argued to function as a contextual background against which specific transport practices are to be understood. In short, while important in their own right, these were felt to lie 'outside' the jurisdictional boundaries of transport governance. To a degree, this reflects the belief that transport is often considered in isolation from other policy domains, with a clear internal mandate and a shared set of professional understandings as to what issues, priorities and instruments are considered appropriate (see Headicar, 2009). In addition, transport is often explicitly or tacitly framed as being relatively subservient to other policy domains and economic sectors. In this sense, rather than a proactive force in its own right, trans-

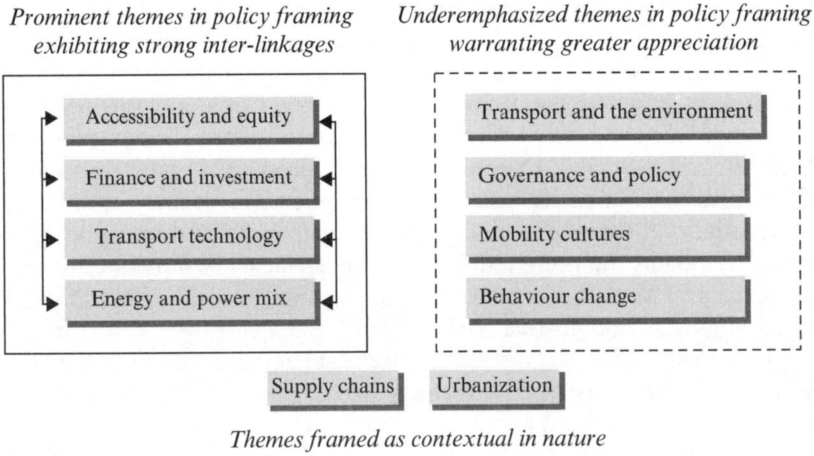


*Figure 12.1* *A simple representation of dominant and underrepresented framings of contemporary mobility*

port is often perceived to be a somewhat passive domain, primarily charged with delivering effective and efficient responses (usually infrastructural) against broader political and economic demands for mobility.

This partly explains why urbanization and supply chains are seen as 'contextual', inevitable socioeconomic developments rather than integral components amenable to intervention through transport governance per se. Urbanization, for example, reflects the relative vitality and economic opportunity present in cities. Naturally, the response of transport policy to urbanization varies spatially and temporally. However, whether or not such responses are intended to manage, curb or indeed foster urbanization varies, but is resolutely seen as a response to an external development. Similarly, supply chains – particularly international logistics networks – may be perceived as removed from the sphere of transport governance in two distinct senses. First, they are private affairs, managed and regulated in accordance with commercial practices. Second, by definition, they invariably span multiple jurisdictions and territories, further reinforcing their status as 'contextual' and beyond the realm of targeted intervention.

The four themes on the left side of Figure 12.1 (in the solid box) arguably exhibit strong inter-linkages and collectively reflect predominant framings and established understandings of the transport system. Accessibility, as a proxy for economic development, is often considered to be of primary importance, in turn providing a rationale for finance and investment, the construction of energy infrastructures and the development of new transport technologies. Collectively, this framing represents what Ney (2009) refers to as the 'efficient mobility' narrative. This narrative is heavily informed by classical economic theory, and treats transport as wholly subservient to the needs of the economy. Essentially this position holds that economic activity demands increasingly time-efficient transport linkages, as time delays represent a curb on productivity and thus negatively influence competiveness. Improving accessibility is thus the ultimate responsibility for transport policy, with private motorized transport generally framed as the preferred focus for investment. In terms of governance, the provision of energy supply and the emergence of novel transport technologies, although usually the responsibility of private actors in the Global North, still enjoy distinct, if often tangential, state support as the vitality of these industries is intricately woven into the efficiency and stability of transport regimes.

In contrast, the four themes on the right side are regularly understood as performing ancillary roles, somewhat tangential to the perceived importance of the other themes. Issues pertaining to the relationship between transport and the environment, while rising to the fore of discourse and debate in policy circles given the salience of anthropogenic climate change, are still understood to rarely factor into bona fide decision-making. This partly reflects the predominance of cost-benefit analyses and related metrics in supporting policy-making, which are notorious for failing to adequately address environmental concerns. The complexities of governance and policy, interestingly, were argued to be under-acknowledged by those working within this area. This was especially stressed where governance regimes at different spatial scales are tasked with collectively addressing particular transport issues.

Behaviour change and mobility cultures were considered the most neglected aspects of contemporary mobility addressed in transport policy and general academic debate. To a significant degree, the academic discipline of transport studies is still rooted in the positivist logic of classical economics and engineering, with so-called 'mobilities' scholarship, although growing in stature, yet to become genuinely mainstream (see Shaw and Hesse, 2010). This is almost certainly all the more true at the interface of policy, consultancy and industry, where the room for discussing the sociocultural nuances of transport practices is slim. This said, interest in behaviour change is growing – reflecting the changing priorities of transport governance discussed above. There was a pronounced feeling, however, that while behaviour change was increasingly recognized as being important, dominant understandings of behavioural change remained relatively crude and under-theorized, with a tendency to rely on long-standing, but much critiqued, assumptions regarding rationality and habit.

## FINAL REMARKS

Having summarized significant points of the preceding ten chapters, this chapter has argued that genuinely understanding patterns and practices of contemporary mobility demands both an appreciation of the various components of transport systems in their own right and a critical awareness of their collective dynamics. In short, it is clear that the exploration of critical inter-theme linkages is vital in order to develop a fuller perspective on contemporary mobility, and to explore its possible evolution. As complex systems, however, modern transport systems are more than the sum of their parts. They cannot be understood as reducible to stable, established and deterministic relationships between variables. Moreover, unlike physical systems, patterns and practices of contemporary mobility represent the collective product of numerous public and private actors' myriad calculated and habitual actions. Importantly, these actions are prone to change on the basis of change itself, leading to pronounced degrees of systemic emergence and unpredictability.

To a degree, the 'partial order' apparent in contemporary transport – manifested in the long-standing dominance of particular technologies, institutions and norms – can appear dangerously seductive to researchers and policy-makers. The prevailing modernist concern for prediction and control draws on specialist knowledge to justify and underpin the formulation of various interventions. However, this partiality means that predictive efforts are inherently limited and such endeavours are intrinsically constrained by narrow conceptual framings and associated gaps in empirical evidence. The dominant, instrumentalist interpretive frame characterizing transport governance is of particular concern. Locked-in to particular ways of viewing mobility, interventions come to be formulated that – while ostensibly mitigating some of the problems associated with carbon-intensive transport – fail to make sufficient progress towards truly sustainable, low carbon mobility. Indeed, when set against the sociotechnical complexities of contemporary mobility, typical academic and policy efforts directed at understanding transport practices appear to be distinctly limited in scope. To be sure, repositories of expertise undoubtedly exist. Yet these rarely interpret the mobility system in a truly comprehensive manner – one that is sensitive to the dynamic interplay existing between the various systemic elements discussed above and in the preceding chapters. As a consequence, transport governance struggles to make headway, with decision-making often involving relatively crude calculations using relatively narrow metrics.

Finally, however, for all the focus on 'managing' transitions to sustainable mobility, it is crucial to recognize that policy and formalized 'interventions' are but one means by which change can occur in systems. Historically, policy and regulatory efforts have certainly played key roles in major transitions, but these have often been secondary in significance compared to the emergence of novel technologies or powerful social movements. Although knowledge produced in relation to the elements discussed in this book will inevitably be partial and transient, therefore, wide-ranging and in-depth analysis must certainly allow for a greater appreciation of the present global mobility system, the complexity of the challenges it faces and the scope for progressive future options.

## NOTE

1. These linkages were identified during an expert workshop hosted by the Transport Studies Unit (TSU), Oxford University, in June 2011 involving TSU transport academics and researchers from the Smith School of Enterprise and the Environment, Oxford University. While additional linkages were explored in small group discussions, those presented here were reported as being of major significance during the workshop's final plenary session.

## REFERENCES

Adey, P., L. Budd and P. Hubbard (2007), 'Flying lessons: exploring the social and cultural geographies of global air travel', *Progress in Human Geography*, **31**(6), 773–91.

Banister, D. (2005), *Unsustainable Transport: City Transport in the New Century*, Abingdon: Routledge.

Banister, D., K. Anderton, D. Bonilla, M. Givoni and T. Schwanen (2011), 'Transport and the environment', *Annual Review of Environment and Resources*, **36**, 247–70.

Cresswell, T. (2006), *On the Move: Mobility in the Modern Western World*, Oxford: Routledge.

Durkheim, E. (1938), *The Rules of Sociological Method*, Glencoe: Free Press.

Geels, F. (2005), 'The dynamics of transitions in socio-technical systems: a multi-level analysis of the transition pathway from horse-drawn carriages to automobiles (1860–1930)', *Technology Analysis and Strategic Management*, **17**, 445–76.

Geyer, R. and S. Rihani (2010), *Complexity and Public Policy*, London: Routledge.

Goodwin, P.B., S. Hallett, F. Kenny and G. Stokes (1991), *Transport: The New Realism. Report to Rees Jeffreys Road Fund*, TSU Working Paper 624, School of Geography and Environment, Oxford University.

Headicar, P. (2009), *Transport Policy and Planning in Great Britain*, London: Routledge.

Jervis, R. (1997), *System Effects: Complexity in Political and Social Life*, Princeton, NJ: Princeton University Press.

Merriman, P. (2007), *Driving Spaces*, Oxford: Blackwell.

Metz, D. (2010), 'Saturation of demand for daily travel', *Transport Reviews*, **30**(5), 659–74.

Millard-Ball, A. and L. Schipper (2011), 'Are we reaching peak travel? Trends in passenger transport in eight industrialized countries', *Transport Reviews*, **31**(3), 357–78.

Miller, D. (2001), *Car Cultures*, Oxford: Berg.

Ney, S. (2009), *Resolving Messy Policy Problems*, London: Earthscan.

O'Connor, A. (2008), 'Tata Nano – world's cheapest new car is unveiled in India', *The Times*, 11 January.

Shaw, J. and M. Hesse (2010), 'Transport, geography and the "new" mobilities', *Transactions of the Institute of British Geographers*, **35**(3), 305–12.

Sovacool, B.K. and R.F. Hirsh (2009), 'Beyond batteries: an examination of the benefits and barriers to plug-in hybrid electric vehicles (PHEVs) and a vehicle-to-grid (V2G) transition', *Energy Policy*, **37**(3), 1095–103.

Urry, J. (2004), 'The "system" of automobility', *Theory, Culture and Society*, **21**(4/5), 25–39.

Urry, J. (2007), *Mobilities*, Cambridge: Polity Press.

# PART II

# Towards low carbon mobility

# 13. Alternative pathways to low carbon mobility

**Moshe Givoni**

## DEFINING AND CONSTRUCTING PATHWAYS

Low carbon mobility is defined as mobility that results in substantially lower levels of carbon consumption and emissions. Here, mobility is defined as the total amount of travel that is undertaken on all forms of transport (Chapter 1), and carbon is used to illustrate the various negative environmental impacts from transport (Chapter 11). As discussed in Part I of the book, and emphasized in Chapter 12, it is the nature of the current transport and mobility system that results in high levels of carbon consumption and emissions, with each of the elements of the system as well as the linkages between them acting to preserve or reinforce high carbon mobility.

The term 'pathway' is commonly used in the futures and transition literature. For example, scenario analysis often involves developing and describing pathways, but there is no single common definition of what a pathway is and usage of the term varies between studies and applications. Geels and Schot (2007) use the term to suggest different ways in which transition might take place (Chapter 14). Foxon et al. (2010) use the notion of pathways to explore how the UK electricity system might evolve if different approaches to managing the sector were applied to achieve a low carbon electricity system. They try to envisage how different external forces and dynamics will affect the approaches taken and create different pathways for the sector, even though they will not necessarily result in the sector meeting the UK carbon emission targets. In other studies a pathway is defined as leading to a particular, desirable future and alternative policy measures are examined to test the different combinations of actions required to move on that path. This is the case in the various Visioning and Backcasting in Transport (VIBAT) studies (Hickman and Banister, 2007; Hickman et al., 2010, 2011). Roadmaps can also be seen as pathways, normally a relatively detailed description of a trajectory of development into the future. Roadmapping as a tool for developing transport policies is suggested, for example, by Tuominen and Ahlqvist (2010). Storylines are also used to describe a pathway to a certain future usually in a more qualitative, often normative and less detailed way (van der Heijden, 2005).

In this chapter pathways are defined in a relatively stylized way. A pathway can be considered as the link between two end points representing a current state, on the one hand, and a future end state, on the other. The two end states are qualitatively different, and some change takes place along the pathway when moving from the current to a future state. When this change is very substantial – a step change, systematic in nature and across multiple domains (for example, technology and infrastructure, policy and regulation, cultural discourses, user practices, knowledge) – it is defined as a transition (Geels, 2002, 2012). The pathways described in this chapter lead from the current high carbon mobility state to a future state of low carbon mobility. A pathway is therefore defined as the route that, if followed, will get us to where we want to be. It is different from the VIBAT approach for two main reasons: the pathway is built from the present to the future and not back from the future to the present, and unlike the VIBAT approach, the pathway might eventually not lead to the desired future. The pathways described in this chapter all lead to low carbon mobility, but to qualitatively different worlds of low carbon mobility – they do not lead to the same future. This is due to the different ways in which low carbon mobility is achieved in each pathway. In this sense, a pathway also serves to provide the direction in which we should turn to, an orientation for policy actions.

The development of the pathways described below followed a three-stage process. First, three distinct approaches to achieve low carbon mobility were identified, based on the understanding of the main mechanism leading to high carbon mobility – namely, the importance of gross domestic product (GDP) growth in policy and the coupling of economic growth, transport volumes and emissions (see Chapter 1). Each approach represents a pathway. Second, the three different pathways were examined in relation to a matrix that includes on one axis the main elements of the mobility and transport system (Chapters 2–11) and three groups of countries on the other axis: developed countries (mainly countries in the Organisation for Economic Co-operation and Development (OECD)), New Industrializing Countries (NICs, for example, China, India, Brazil) and developing countries. Third, a workshop then took place[1] to interrogate the pathways in the context of the matrix. The main issues raised with respect to each pathway are included in the description of the pathways below.

The pathways are built, and described, around four main elements: emissions – carbon is used to represent the negative environmental impact of mobility activities; mobility – the total volume of transport; growth – the increase in GDP and economic factors; and lifestyle – a term to capture a wider set of social norms and values that is a manifestation of everyday life and choices about, amongst other things, mobility. In the present high carbon mobility system emissions, mobility and growth are all closely interlinked (Chapter 1). High carbon mobility can be said to be mainly the result of three factors: the centrality of economic growth in mainstream political, policy and socio-economic discourses; the high positive correlation between economic growth and mobility levels; and the high positive correlation between mobility levels and carbon emissions (as well as other gases and pollutants). Therefore, growth, mobility and emissions need to be separated to achieve low carbon mobility, unless current lifestyles are changed. Thus, the notion of decoupling – separating or detaching these elements – is the tool or mechanism through which low carbon mobility can be achieved. What is decoupled from what differentiates the pathways described in the next section and in Figure 13.1. It can be argued that without any change in these linkages the 'business as usual' (BAU) pathway will lead to a future of even higher carbon mobility with all three elements, economic growth, mobility and emissions, higher than today (Figure 13.1). There is evidence however that the BAU trajectory is already different (see the final section).

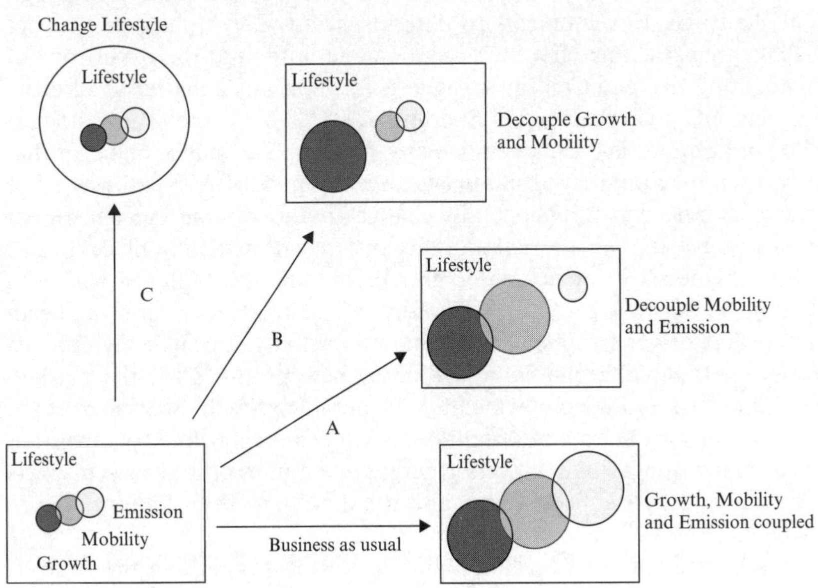


*Figure 13.1 Alternative pathways to low carbon mobility*

The pathways are first presented as three alternatives with no hierarchy between them, although some commentary is given on likely risks, drawbacks and challenges associated with each. There is also no discussion in this chapter on how these pathways might be selected, realized or play out in real life when influenced by various processes and dynamics. The aim here is mainly to present the alternatives. After a description of each of the pathways in the next section, they are summarized and discussed in the following section. The concluding section offers some commentary on the pathways with respect to the current debate on and policies for low carbon mobility.

## ALTERNATIVE PATHWAYS TO LOW CARBON MOBILITY

Figure 13.1 illustrates the three pathways that are described and discussed in detail below.

### Pathway A – Mobility with Lower Emissions

A straightforward way to achieve low carbon mobility is to decarbonize transport via technological developments. Decoupling transport from emissions to achieve low carbon mobility is thus mainly a technological challenge. Developments to date of various Zero Emission Vehicles (ZEVs) suggest this challenge is attainable, although the experience of mandating ZEVs in California suggests it is not only a matter of agreeing on the policy (Collantes and Sperling, 2008). This is the 'Technological Fix' option that is attractive for many reasons. The main one being that it does not require a substantial change from the BAU pathway since there is no need to fundamentally change anything in our current way of living, suggesting the current mobility system can remain in place. In this pathway mobility and economic growth are still closely linked and both continue to grow. Moreover, this pathway is also attractive as it can be an important driver for enhancing economic activity, through investments in green mobility technology leading to new products for the market, and thus facilitating employment and economic growth (Schwanen et al., 2011). It can also be an important driver for enhancing global competition and innovation as new players emerge in the market and new markets are created, for example, China and the development of battery electric vehicles (BEVs).

Under this pathway mobility continues to rise mainly in the form of increasing air transport volumes (for passengers and goods) and shipping volumes (goods only). Surface transport also continues to grow in volumes, but more in the NICs where levels of mobility (per capita) are still relatively low, linked to a relatively low GDP per capita, but where economic growth rates are high. The growth in surface transport in those countries is realized through massive expansion of transport infrastructure networks, including road, rail and high-speed rail. In the NICs the increasing levels of car ownership will be another major driver in the growth of mobility. In contrast, in the developed countries, like the OECD countries, capacity constraints on the existing transport infrastructure (mainly road) and already high levels of car ownership will slow the growth in transport. It may even halt this growth, a phenomena that is already observed in some countries (Metz, 2010; Millard-Ball and Schipper, 2011). For developing countries, there is uncertainty about the prospects for substantial, long-term economic growth and therefore uncertainty over growth rates in mobility and emissions. Yet, it might be that technological developments, such as a new generation of cars powered by renewable fuel sources, might mean that decoupling could have the most impact here, in terms of a reduction in emissions per passenger or tonne-km, since the current performance of the transport vehicle fleet is relatively poor. Alternatively, if lack of economic growth (or other reasons) means those countries cannot afford the new technologies, decoupling might not take place.

The same set of values currently defining our lifestyles and our mobility system drives this pathway. Speed continues to be central, and therefore travel time and not distance is important. Thus, farther is not necessarily inferior to closer, but depends on travel time. What is different under this pathway and what very much motivates its development is increasing environmental awareness and a real concern with the prospects of an anthropologic climate change and other transport-related environmental impacts, such as air pollution and noise in cities. Pathway A and the approach it advocates to address the low carbon mobility challenge assumes that environmental and economic objectives do not necessarily contradict each other, and can be pursued in tandem. This approach has been labelled 'ecological modernization' (see Schwanen et al., 2011 for a brief discussion) and is represented in terms like 'green growth' and 'green economy' and indeed 'low carbon economy', which have been increasingly used since the recent financial crisis began (see Jänicke, 2012 for a discussion and review of relevant policy documents).

This pathway does not require a systematic change, a 'transition' as defined above, but a much more subtle change where research and development (R&D) efforts in surface transport, and car in particular, are focused largely on efficiency and not on performance of vehicles, or performance is focused on fuel efficiency and not on speed, carrying capacity (size) and energy-consuming accessories.[2] Such efforts would not only be motivated by environmental issues but also, at times even more, concern about high oil prices, their economic implications and concerns about peak oil. Similarly in air transport, fuel efficiency rather than speed and range would drive the development of the latest generation of aircraft, and this objective also fits well with increasing public concerns about air transport, its impact on the environment and the industry's green image. Likewise in shipping, fuel efficiency would drive technological innovations.

This attractive pathway to low carbon mobility does, however, have several limitations. For example, electric cars, which best typify this pathway, might result in low carbon transport, but might simply shift carbon emissions elsewhere, in this case to the energy sector. This will mean a substantial increase in demand for electricity, intensifying the challenge of decarbonizing electricity. Furthermore, the effect of technological development, and also of other policies, is often eroded, even eliminated in some instances, due to rebound effects (Druckman et al., 2011). This pathway also further locks us into current mobility patterns that have other adverse effects beside carbon emissions, air pollution and noise problems. If we achieve low carbon mobility by decarbonizing transport via technological developments, urban sprawl will continue and cities might grow even larger than those levels resulting from population growth, making distances greater and locking us into longer commuting distances and more time on the transport network. The side effects here include greater levels of car use, more road accidents and continued sedentary lifestyle, obesity and other related health problems (especially in OECD countries and increasingly in NICs). To cope with the increasing demand for transport, there will be further pressure to expand the (road) transport network and this in turn will place additional demands on the remaining open, green spaces, and will likely affect ecosystems and biodiversity. The environment will remain an issue of concern. Although congestion will not anymore be a 'carbon' problem it will nevertheless remain a problem, and efforts to reduce congestion will further motivate expansion of the infrastructure, even if this leads to the generation of new demand for travel (Goodwin, 1996).

Pathway A allows us to reach low carbon mobility by simply inventing new vehicles, and especially vehicles that do not consume oil. The rest (more or less) can remain unchanged and this is what makes this pathway the easy choice. However, it carries the risks that (a) the low carbon mobility target will not be attained, for example, because we failed to find an alternative fuel for aircraft and shipping and/or we failed to fully decarbonize electricity, and (b) the other problems associated with the current mobility and transport system will remain and intensify. In this pathway the agency lies mainly with the private sector to come up with the technological solutions, and ones that must also be commercially viable and suitable for mass production. The public sector also has an important role to play in supporting and incentivizing (through pricing, subsidies and other economic tools) the private sector and by adopting supporting policies, like widespread adoption of Low Emission Zones (LEZs). This pathway is not dependent on changing public attitudes and social norms, and so does not involve the public directly, but it is dependent on the public adopting the new technologies, and the faster the better. As such, mobility culture and travel behaviour do not change.

**Pathway B – Growth with Lower Mobility**

A second possibility of achieving low carbon mobility is to decouple economic growth from mobility. Whether larger volumes of transport support and even induce economic growth or are just the result of economic growth is still an open debate, but the two are certainly very closely linked (Chapters 9 and 10). If we can maintain or increase rates of economic growth without a corresponding increase in mobility or even a decrease, we could achieve low carbon mobility. In this case there is no need in principle to decouple transport from emissions and the technological fix is less needed. We can maintain our current lifestyles, but this pathway would require changes in everyday practices since mobility stops growing or even declines. There are clear elements of a transition in this pathway.

This form of decoupling has long been seen as a way to reconcile economic and environmental concerns, and it is also the way the use of the term 'decoupling' has been applied in transport research and policy. Based on the work of Vehmas et al. (2003), Tapio (2005) defines several categories of decoupling, essentially looking at whether GDP and transport volumes each continue to grow or not, and at what rate relative to each other. This research on 15 European Union (EU) countries (1970 and 2001) found that all have experienced growth in GDP and transport. In some countries transport grew even faster than GDP and in others GDP grew at a faster rate than transport. The two situations described above illustrate two forms of 'relative decoupling', where the concern is over the relative changes in GDP and transport. Pathway B, however, requires an 'absolute decoupling', where transport growth is negative (that is, less travel) and GDP growth is positive. Tapio (2005) continues the analysis by applying the same categories of decoupling to the growth in GDP and in carbon dioxide ($CO_2$) emissions with similar results. All 15 EU countries experienced no absolute decoupling but different degrees of relative decoupling (see also discussion below, and Tapio et al., 2007). These findings reinforce the current coupling of growth, mobility and emissions as illustrated in Figure 13.1 for the BAU pathway. Similar relative decoupling was also found by McKinnon (2007) for economic growth and growth in freight transport in the UK in the years 1997 to 2004.

Pathway B, in essence, heralds a shift from standard processes of globalization to a model of 'glocalization', where self-contained local/regional markets mean that the production, distribution and consumption of goods and services are performed across much shorter distances. Glocalization entails reducing the demand, and need, for long-distance physical movement of people and goods. It requires dematerialization of transport, such that increasingly knowledge is 'transported' while materials are sourced locally – what is often referred to as transporting electrons rather than molecules. In a glocalized economy, 'just-in-time' as well as long (transcontinental) supply chains (Chapter 10) are no longer seen as viable business models, but there is still room for businesses to be global and there is room for multinational global companies and corporations, hence the reference to 'global' in 'glocalization'. Under glocalization, and Pathway B, 'relocalization' is the motto for new, alternative business models as advocated by Wells and Nieuwenhuis (2006). Similarly, localization is promoted across various activities. Smaller 'activity centres' (for example, of employment, commerce, production, education and leisure) are more evenly distributed across space and they will gain importance, and become more commercially viable and attractive, certainly for customers but also for suppliers as the benefits of economies of scale, largely supported today by low transportation costs, will become substantially less important. As Wells and Nieuwenhuis (2006) note, economies of scale under the model of relocalization will be moved up the supply chain and away from the production line (or 'assembly line' in Wells's and Nieuwenhuis's words as they mainly refer to production of cars) and this will support the substitution of moving electrons rather than molecules. Instead of transporting end products long distances, knowledge and know-how will be mobilized so that local production can take place. Phrases like 'think global act local' with respect to the economy, not only the environment, illustrate this pathway. What counts as 'local' will naturally change depending on activity, product, context and so on. With respect to economic activity, or that of production of goods, Wells and Nieuwenhuis (2006) see 'local' as being primarily national.

In Pathway B economy of scale becomes diseconomy of scale. First, the current diseconomies of scale are recognized, largely associated with overproduction as, for example, noted by Wells and Nieuwenhuis (2006, p. 94) for one industry: 'The automotive industry shows symptoms of over-production. These include high rates of discounting for new products, high rates of product value depreciation, ownership rates that greatly exceed the ability of people to use the product and premature scrapping of the product. Over-production, a strong possibility with capital-intensive plant, is undesirable because it constitutes a waste of resources and unnecessary pollution.' Second, and likely to be more important to facilitate a move from a globalization to a glocalization model, the full economic as well as environmental cost of transport is included in the prices of all products, and this in turn requires pricing of carbon and other environmental harms at much higher levels than are prevalent today. For example, the concept of food-miles[3] will be fully adopted, and this would apply to the costs of distribution for all consumption goods and the cost of a product-mile would be much higher than at present. There is, therefore, much more concern for and regard to the 'value' of what is moved around as transport costs are much higher.

In passenger transport time is still seen as having a negative utility, but efforts would concentrate on reducing distance rather than on increasing speed. And the notion of travel time reliability (Banister, 2008) would gain more weight over travel time savings in driving transport service providers' success and in ensuring customers' satisfaction. Effective use of time while travelling now becomes a key element of travel, and this includes increasing the opportunities to either work or engage in various leisure activities while travelling (reading, conversing, resting and so on). Attention would need to be given to in-vehicle and out-of-vehicle travel environments, including higher levels of comfort, while better utilizing information and communications technologies (ICT) possibilities. This strategy is likely to be central in facilitating a mode shift from car travel to public transport. Although Pathway B does not explicitly advocate for such policy, since a shift from car to rail on inter-city journeys can in principle result in longer travel distances and more travel, it can be an important component of this pathway, especially with respect to urban transport. This means that integrated transport policies (Givoni and Banister, 2010) are central in this pathway.

Since reducing transport distances is key, this pathway advocates the densification of cities, which will support a shift to non-motorized transport, primarily walking and cycling, and the use of public transport. Pathway B requires the deployment of ICT to substitute for, and not complement, transport. Thus home-working, teleworking and teleconferences would be adopted with actual face-to-face meetings minimized and increasingly becoming the exception.

In this pathway the public sector, mainly national governments, has the role of the principal agency to advance it, mainly through incentivizing the private sector to adopt a different business model while imposing restrictions on current ones that contribute to high carbon mobility.

Governments can support local and smaller businesses, encourage public support for these businesses (for example, buy locally campaigns), and they can penalize, largely through pricing mechanisms, any transport-intensive business models. The public in this pathway has a much greater role to play, as it can support and further push the public and private sectors alike to progress in this direction.

Pathway B brings about low carbon mobility by essentially keeping the current transport system but changing the economic approach, or economic growth model, to one that is less reliant on transport of goods and people over long distances – a model of glocaliztion. Economic growth is still the major concern in national and regional policies, but this aim is achieved through different means to those in use today. Under this pathway the economy will grow, but with substantially less transport, so there is less concern about increasing carbon from transport and hence less motivation, or urgency, to decouple mobility from emissions.

In this pathway change is mainly driven by new initiatives in governance and policy, but the mobility culture and travel behaviour also change substantially – even if lifestyle remain similar in general – affecting all other elements of the mobility and transport system. For example, investments are less directed towards physical (transport) infrastructure, and technology focuses more on substituting travel rather than facilitating it. Concerns over energy security and a preference for less reliance on imports and exports in times of geo-political uncertainty might also encourage the adoption of Pathway B. This can undermine the current growth strategy of the NICs, which relies heavily on relatively cheap transport that facilitates trade, and it will force them to base economic growth on local markets or to move away from the BAU pathway, even if initially this is not their choice. For developing countries this pathway represents a significant challenge, but it might also be an opportunity to independently determine and shape their economies, so that it is based more on local considerations than on the fortunes of global trade and other countries.

## Pathway C – Changing Lifestyle

A third pathway to low carbon mobility is one that requires the most radical change (a full transition) through adopting a different lifestyle based on a different set of values. These new values are only discussed in general terms, as the focus is on the implications with respect to mobility. The general prevailing norms are 'growth is good', 'more is better' and likewise 'faster is better' and this means that time has a high value – 'time is money'. Slow is seen as a waste of time and money, and it is therefore inefficient. In today's world (largely the Western world), an accelerating pace of life is translated through a variety of ways to higher levels of mobility, especially since our life is nowadays organized around a much larger spatial space –thanks to development and investment in transport technology and infrastructure. While this does not hold across all society and is not necessarily true for all individuals, the implications nevertheless of increased mobility are felt across society. A paradigm shift towards 'less can be better' that in turn results in slowing the pace of life and 'squeezing' less into the 24-hour day will result in lower mobility as shorter distances will need to be overcome. Perhaps also less time will be devoted to physically moving from one place to another.

There are some similarities here with Pathway B, as localism is highly valued, but the change is more fundamental. First, the change is driven by a different set of social values and policy priorities that dictates a different economic model, not a similar set of values and policies (which prioritize economic growth), as is the case in Pathway B. 'Prosperity without growth' (Jackson, 2009) or 'Rethinking growth' is perhaps the most important change that drives Pathway C. Rather than economic growth as defined at present, namely, the increase in GDP per capita (mainly considered at a national level, not the global or individual levels), as the key driver in the BAU pathway and both Pathways A and B, the focus here is on growth, which is qualitatively different and encompasses a broader set of indicators than GDP. In Pathway C the limitations of GDP as an indicator for prosperity are accepted (thus resolving the GDP paradox, van den Bergh, 2009), and wellbeing more generally becomes the main concern for policy, with due attention being given to health, education and social care. Likewise, equality takes precedence over just higher averages of wealth, which often result in increasing the gaps between different parts of society.

There are many different ways to define the term wellbeing, largely depending on the discipline within which the term is used, as illustrated by the ten different approaches identified by Nordbakke and Schwanen (forthcoming). Pathway C is about moving from a definition of wellbeing as a utilitarian conception rooted in economic theory, where income is the proxy for wellbeing, to a definition that aligns wellbeing with the capability approach associated with the work of Amartya Sen and Martha Nussbaum (Beyazit, 2011). The capability approach describes capability as the 'ability to achieve various combinations of functionings that we can compare and judge against each other in terms of what we have reason to value' (Sen, 2009, p. 233 in Beyazit, 2011, p. 121) and it takes a much broader perspective on development and wellbeing. It also encompasses the notion of 'basic capabilities', described as the 'ability to move, to meet one's nutritional requirements, the wherewithal to be clothed and sheltered, the power to participate in the social life of the community' (Sen,

1980, p. 367 in Beyazit, 2011, p. 120). It therefore allows a broader consideration of poverty accounting for multiple criteria based on 'identifying a minimal combination of basic capabilities' (Sen, 1993, p. 41) rather than considering minimum income only. Likewise, the capability approach allows a much broader consideration of (in)equality. In Pathway C levels of health, education and social care are the main policy priorities and not growth in GDP.

Society in Pathway C is much more egalitarian and issues of social justice are central. This has an enormous effect on mobility and how the transport system is organized. The perception of public transport, especially when it is compared with private transport, will change dramatically and be favoured. The individualistic, even 'selfish' character of car use and the damage cars can inflict, especially in terms of road casualties and injuries, could lead to severe restrictions on car use, and not only for environmental reasons (see Pooley, 2010 for a discussion of such a counterfactual scenario for the development of the UK transport system). The capability approach also allows a different consideration of accessibility in which financial capability to access places is only one dimension of accessibility (Chapter 3). Importantly, it means evaluation of transport policies and investments goes beyond cost-benefit analysis where the monetary value of various costs and benefits determine whether action (for example, investment) is worthwhile (Beyazit, 2011).

The key to Pathway C is a fundamental shift in societal values that results in alternative forms of production and consumption, and perhaps also different forms of ownership (shared rather than private, for example), and a change of lifestyle. This shift will lead to a decline in consumerism and a reconfiguration of global production chains. As Figure 13.1 illustrates, there is in principle no need for decoupling in Pathway C. There is no need to decouple mobility from emissions or growth from mobility, as the new way of living is different, one where economic growth, and therefore mobility and emissions, is modest and different.

In 1854 Thomas Cook declared as the slogan for the time: 'To remain stationary in these times of change, when all the world is on the move, would be a crime' (Urry, 2007, p. 14 quoted in Brendon, 1991, p. 65), and this seems to still hold today. In contrast, 'It is a crime to move endlessly over more distances to ever more places' might be the slogan representing Pathway C. While 'slower is better' might not be the norm, 'closer is better' certainly is, and thus reasonable travel time takes precedent over minimum travel time as suggested by Banister (2008). In Pathway C commuting two hours per day or flying around the world (in aggregate) once a year will be perceived as unreasonable, and for different reasons – too expensive, time-consuming or simply unacceptable (for environmental or social reasons).

Such a change in social norms, and in particular a much greater priority allocated to slower travel, will lead to change in business models and to economic activity that is realigned across smaller scales, spatially and in size. The business models argued for by Wells and Nieuwenhuis (2006) and described in Pathway B are similarly applicable here, and maybe even more so. The reason is that they are largely derived from principles set by the economist Ernst Friedrich Schumacher and aim not only to make the economy 'better' but also account for other, social concerns including issues of equality (see Wells and Nieuwenhuis, 2006, p. 95). For example, local tourism takes precedent here over international tourism with virtual international tourism substituting for some of the latter. Overall, there is less emphasis on consumption in all its forms. In transport overcoming or shrinking distance is not a priority and therefore there is less investment in long-distance travel while in cities the full benefits (including health and social) of walking and cycling are recognized. Thus, cities are designed and organized accordingly, with land use determined by reasonable walking and cycling distances (these distances are likely to be greater than today; Tight et al., 2011).

Pathway C is neither about the environment nor about reconciling economic and environmental concerns; it is about living differently, partially for environmental reasons and concerns for the environment. Thus, and in principle, even if transport is emission-free and relies on renewable energy, it will be seen as a 'crime' to move endlessly, due to the social implications of being 'constantly' on the move. Likewise, even if speed means distance can be overcome with a relatively small time penalty and nearer does not necessarily mean less travel time, proximity will still be preferred. Pathway C might be seen as representing a step backwards, but it does not need to be. Also in Pathway C technology can still be harnessed and technological development encouraged, but to serve and not lead a socio-technical transition. Furthermore, the R&D efforts under Pathway C are still important and are concentrated outside the transport sector (more than in Pathway B and unlike in Pathway A) – mainly to improve and develop substitutes for transport. Thus, the potential for innovative and radical new technologies and their utilization in everyday life to substitute for the current transport infrastructure network is probably most prominent in Pathway C.

Pathway C is likely to be attractive especially in places where economic prosperity across much of the population has already been achieved, such as in the OECD countries. But for the NICs and developing countries this pathway represents a dilemma as it suggests that the standard of materialistic living enjoyed across OECD countries might not be attainable. However, under this pathway developing countries are not necessarily underdeveloped, although wellbeing undoubtedly needs to be improved if health, educational standards and social care are considered. Yet, if the efforts are concentrated at improving wellbeing, by adopting the capability approach conceptualization of wellbeing, rather than a narrow interpretation of an economic standard of living, the prospects for improvements might be higher. In this respect Pathway C might be an opportunity to unlock existing inequalities across countries and regions.

Agency in this pathway lies mainly with the public, and they must drive social change and in turn influence the political system and the private sector. As indicated above, this change also affects mobility cultures and travel behaviour and by this, mobility levels. Accessibility will be another important dimension of wellbeing and a policy concern, and in a different way than it is today; the accessibility of people and not the accessibility of places as such (for example, by car) will become the key to an inclusive and fair society.

## FROM THREE TO NUMEROUS PATHWAYS

A summary of the three pathways presented above, which represent the main different options to achieve low carbon mobility, is shown in Table 13.1. Considering the socio-technical transition required to achieve low carbon mobility (see Chapter 14), it can be said that Pathway A is led by a technological transition while Pathway C by a social one, with Pathway B in between. The scale of change envisaged, with respect to the current mobility and transport system, is largest in Pathway C, making it perhaps the hardest to achieve, but the one with the best prospects to sustain low carbon mobility. Pathway B would also require a substantial

*Table 13.1 Three alternative pathways to low carbon mobility*

| | Pathway A | Pathway B | Pathway C |
|---|---|---|---|
| Name | The Technological Fix | Glocalization | Rethinking growth |
| Description | Mobility with less emissions | Growth with less mobility | Changing lifestyle |
| Decoupling | Mobility from emissions | Economic growth from transport | Wellbeing from economic growth |
| Dominant transport policy | Improve | Shift | Avoid |

change in the way of living and in 'thinking' when it comes to the organization of production and consumption and the resultant transport, but it does not challenge the prevailing principles leading to high carbon mobility, as Pathway C does. In essence for Pathway B, the supremacy of economic growth in determining society's wellbeing is not questioned, making it easier to adopt than Pathway C. Pathway A might appear in this respect attractive, since it does not require much change other than funnelling (technological) development efforts towards the production of low emission transport, or even better, ZEVs. However, the prospects for Pathway A to sustain low carbon mobility might not be promising, since it is likely to result in relative and not absolute decoupling and the shift of carbon elsewhere. It would be very difficult (if not impossible) to achieve low carbon mobility through this pathway.

Chapter 1 made reference to three categories of transport policies used to reduce the carbon impact of transport: avoid, shift and improve.[4] In some sense and in policy terms, Pathway C relies on avoid, Pathway B on shift and Pathway A on improve policies to achieve low carbon mobility. While all three types of policies are necessary, the key to moving from high to low carbon mobility lies in adopting first avoid policies, then shift policies and lastly improve policies. Improve policies and to a degree shift policies are more likely to only lead to relative decoupling of transport and emissions. This means that while per unit of transport (for example, passenger-km) emissions are reduced, forces acting to increase demand and the possible rebound effects all lead to an increase in the total level of transport and therefore emissions (although probably by less than if the improve policies had not been introduced). Absolute decoupling of transport and emissions occurs when the total volume of transport increases but the total volume of emissions declines. Avoid policies result in less transport and therefore less emissions, thus the coupling of transport and emissions is of less concern.

The way to low carbon mobility does not necessarily go through one of the above pathways but can include any combinations of them over time or some 'middle' or mix of pathways at any given time. Furthermore, it is most likely that a combination of all three pathways will create the conditions for actually achieving low carbon mobility and it might be that different regions will adopt different pathways or a different mix of pathways. In this respect, the pathways are not mutually exclusive, and there are many pathways of pathways. Each would represent a different combination of the above pathways, and even others that have not been considered here. Moving from high to low carbon mobility depends on circumstances that are different, and most likely unique to a place, so different pathways should be selected and followed. However, the sequence in which Pathways

A, B and C, or ingredients of them, are selected and implemented will have implications for the likelihood of achieving low carbon mobility.

This means that to adopt Pathway C as an alternative to the BAU pathway, there might be a need to move first to Pathway A and then B before embarking on Pathway C. However, Pathway A is likely to further lock society into the current automobility system, which typifies the current high carbon mobility system. This means that if Pathway A is adopted, there might not be enough momentum and urgency to move towards B and particularly C. This might result from the perception that the carbon problem of transport has been solved, although other problems associated with high mobility will remain and even intensify. Furthermore, as noted, the prospects for Pathway A to sustain low carbon mobility might not be promising, since it is likely to result in relative and not absolute decoupling and the shift of carbon elsewhere. The prospects for sustaining low carbon mobility are much higher if Pathway B is adopted successfully, but there is a risk of 'slipping' back to Pathway A or BAU, since path dependency means the current economic approach is likely to be a better approach to achieving growth. Pathway C is more stable, with respect to low carbon mobility, as the motivation for changing 'back' will be weak. Path dependencies and lock-in effects generated within and as a result of each pathway will more likely work in favour of keeping low carbon mobility in Pathway C than in Pathways A and B, where such effects might apply pressure to return to the current high carbon mobility.

A central assumption underpinning the BAU and Pathways A and B is that economic growth takes place under all alternatives. But what are the implications of the current financial crisis continuing and economic growth not taking place for a long time, despite political efforts? In principle, and due to the way the pathways are constructed, the result will be a move towards lower carbon mobility. Without economic growth and without resources to invest in transport infrastructure and mobility technology, and, more important, with a need for new business models that are not primarily aimed at generating growth and maximizing profit (but maybe to maintain production and to support the current workforce), Pathway C might emerge as the result.

## WHERE ARE WE NOW WITH RESPECT TO THE PATHWAYS?

Research by Finel and Tapio (2012) covering the time period 1975 to 2005 and 137 countries found that at the global level, most countries experienced growth in both GDP and volume of $CO_2$ from transport (see also Tapio et al., 2011). The results are similar to the earlier analysis on the EU-15 countries referred to earlier (Tapio, 2005; Tapio et al., 2007). Although mobility levels are not reported (Finel and Tapio 2012), the analysis implicitly examines the links between economic growth, mobility levels and the resultant $CO_2$ emissions, and it illustrates the close links between these three variables that typifies the BAU pathway. In some countries GDP increased more than $CO_2$ emissions from transport – showing a degree of decoupling, but only relative decoupling representing some shift towards Pathways A or B. The analysis also shows that some countries, 21 in number, did experience the type of decoupling advocated in Pathway B over part of the examined period – GDP grew but the volume of $CO_2$ emissions declined. Such a situation can represent either Pathway A or Pathway B as decoupling could be due to decoupling growth from mobility, mobility from emissions or a mix of both. Furthermore, some countries, although a relatively small group, experienced both a reduction in GDP (negative growth) and in $CO_2$ emissions from transport, but never across the whole time period examined. This, however, does not indicate that these countries were, even for a limited period, following Pathway C, even if the outcome is similar (reduction in both growth and emissions and likely also mobility). The reason is that reduction in growth was not of choice and was considered undesirable, as it reflected a period of recession.

In many respects, it can be argued that the need to address the carbon problem of transport has been acknowledged and a decision has been made to advance towards low carbon mobility. This comment is supported by the numerous studies on sustainability and/or climate change and transport made by different stakeholders (for example, IEA, 2008; McKinsey & Company, 2009; Netherlands Environmental Assessment Agency and Stockholm Resilience Centre, 2009; Skinner et al., 2010, amongst many). While the different notions of change advocated by each of the pathways are present in the policy recommendations suggested in the different studies, there is arguably too much reliance placed on Pathway A – the Technological Fix (also in academia; see, for example, Sperling and Gordon, 2009). Schwanen et al. (2011) in their review of scientific research about climate change arrive at a similar conclusion.

Arguably, in many respects a choice has been made and the BAU path already follows Pathway A, although it is too early to say if we are on Pathway A. Much policy thinking and actions, with the encouragement of the private sector, is directed towards 'greening' of the economy and, an important part of it, 'greening' of transport. It seems that the ecological modernization approach is gaining momentum. As evidence from Chapter 7 shows, there are already many promising technological developments that can further reduce emissions from transport, especially carbon emissions. At present, policy action is opposite to that advocated herein, with avoid policies getting minimal support and backing compared to shift policies, which are increasing in importance (for example, a shift from air to rail and from private to public transport), but neither of these are so central as the improve policies (see also Unruh, 2000; Banister et al., 2011). A limitation with the current BAU pathway that follows Pathway A, which seems to be the choice at present, is that the efforts to decarbonize transport goes alongside continuing efforts to make transport more efficient, in terms of minimizing travel time and increasing flows, largely to accommodate and facilitate economic growth, thus still maintaining high carbon mobility in practice despite efforts to move towards low carbon mobility. While success in terms of reducing emissions at the point of production has been large, there has not been much success yet in reducing emissions at source (Banister et al., 2011). With issues of rebound effects, and the transfer of carbon emissions offshore or to a different sector well known, and with the lack of significant reduction in absolute emissions from transport it is unlikely that Pathway A, as the BAU pathway, will provide a real step forward towards low carbon mobility.

Current policy efforts to advance low carbon mobility are too much focused on the potential for change inside the transport system, and the more important external factors that determine and drive the demand for transport have been generally ignored. These are the main factors that explain current mobility levels and they are not given due attention. A broader perspective is required and an understanding that change should first take place outside the transport and mobility system, and that this system needs to adapt to that change for low carbon mobility to become realistic. Moreover, there is an urgent need for recognition that the problem of high carbon mobility is primarily a social one and therefore a technological solution cannot provide the solution, or must at best be only part of the solution.

The three pathways, or alternatives, presented above must guide political and policy thinking and courses of actions. While the options are more or less clear, steering towards a new direction is a challenge in itself, as the next chapter explains. Chapter 14 shows that what might encourage or cause a shift from the current BAU/Pathway A choice to Pathways B or C is a change in the 'landscape'. For example, if and when issues of climate change and peak oil become more prominent and provide a more realistic challenge to the public (and not a future, potential challenge, the Giddens Paradox;[5] Giddens, 2009), the conditions for moving to a different pathway will emerge. In many ways, the move towards Pathway A is unwelcome as it does not really represent a change in policy; it only replicates BAU with a higher level of technological change. It may be doubly unwelcome as it delays the need to make more fundamental decisions about real change and the need for either an absolute decoupling of mobility (and carbon) from economic growth or the redefinition of economic growth itself as the primary driver of global wellbeing. Lastly, there is a risk in ignoring the fact that despite substantial technological development to date, which has substantially reduced emissions (of $CO_2$ and other pollutants) per unit of transport (that is, tone or passenger-km), the environmental impact of transport is steadily increasing.

## ACKNOWLEDGEMENT

This chapter is based on contributions from participants of a workshop hosted by the Transport Studies Unit in July 2011 as part of the Low Carbon Mobility project, at Wolfson College, Oxford University. The author would like to thank David Banister and Tim Schwanen for comments and suggestions on earlier drafts of this chapter.

## NOTES

1. Hosted by the Transport Studies Unit in July 2011 as part of the Low Carbon Mobility project.
2. Arguably, for the car industry and car manufacturers especially, the technological change, like the move from the internal combustion engine fuelled by oil to an electric engine, will amount to a transition.
3. All the transport costs (per mile) associated with the consumption of food products, and including, for example, those associated with the cultivation of crops and plants and their transport to the market.
4. '"Avoid" suggests reducing the demand for transport and thus reducing its volume. Some volume of transport will always be required and this volume therefore needs to be "shifted" to the most environmentally efficient modes. Furthermore, the environmental performance of these modes needs to be further "improved", so that carbon (and other emissions) per unit of necessary transport by the most efficient mode of transport will be further reduced and minimized' (Chapter 1).
5. The Giddens Paradox refers to a situation where people, or society, will not act to prevent a catastrophe or disaster (like climate change) even if they are aware of it approaching, until it is too late to avoid it. This is because the approaching catastrophe does not affect everyday life and does not pose an immediate problem.

## REFERENCES

Banister, D. (2008), 'The sustainable mobility paradigm', *Transport Policy*, **15**, 73–80.

Banister, D., K. Anderton, D. Bonilla, M. Givoni and T. Schwanen (2011), 'Transport and the environment', *Annual Review of Environment and Resources*, **36**, 247–70.

Beyazit, E. (2011), 'Evaluating social justice in transport: lessons to be learned from the Capability Approach', *Transport Reviews*, **31**(1), 117–34.

Brendon, P. (1991), *Thomas Cook: 150 Years of Popular Tourism*, London: Secker and Warburg.

Collantes, G. and D. Sperling (2008), 'The origin of California's zero emission vehicle mandate', *Transportation Research Part A*, **42**(10), 1302–13.

Druckman, A., M. Chitnis, S. Sorrell and T. Jackson (2011), 'Missing carbon reductions? Exploring rebound and backfire effects in UK households', *Energy Policy*, **39**(6), 3572–81.

Finel, N. and P. Tapio (2012), *Decoupling Transport $CO_2$ from GDP*, Finland Futures Research Centre, University of Turku.

Foxon, T.J., G.P. Hammond and P.J.G. Pearson (2010), 'Developing transition pathways for a low carbon electricity system in the UK', *Technological Forecasting and Social Change*, **77**(8), 1203–13.

Geels, F.W. (2002), 'Technological transitions as evolutionary reconfiguration processes: a multi-level perspective and a case-study', *Research Policy*, **31**, 1257–74.

Geels, F.W (2012), 'A socio-technical analysis of low-carbon transitions: introducing the multi-level perspective into transport', *Journal of Transport Geography*, **24**, 471–82.

Geels, F.W. and J. Schot (2007), 'Typology of sociotechnical transition pathways', *Research Policy*, **36**, 399–417.

Giddens, A. (2009), *The Politics of Climate Change*, Cambridge: Polity Press.

Givoni, M. and D. Banister (eds) (2010), *Integrated Transport: From Policy to Practice*, London: Routledge.

Goodwin, P.B. (1996), 'Empirical evidence on induced traffic: a review and synthesis', *Transportation*, **23**, 35–54.

Hickman, R. and D. Banister (2007), 'Looking over the horizon: transport and reduced $CO_2$ emissions in the UK by 2030', *Transport Policy*, **14**(5), 377–87.

Hickman, R., O. Ashiru and D. Banister (2010), 'Transport and climate change: simulating the options for carbon reduction in London', *Transport Policy*, **17**(2), 110–25.

Hickman, R., O. Ashiru and D. Banister (2011), 'Transitions to low carbon transport futures: strategic conversations from London and Delhi', *Journal of Transport Geography*, **19**(6), 1553–62.

IEA (International Energy Agency) (2008), *Energy Technology Perspectives – Scenarios and Strategies to 2050*, Paris: OECD and IEA.

Jackson, T. (2009), *Prosperity without Growth: Economics for a Finite Planet*, London: Earthscan.

Jänicke, M. (2012), 'Green growth: from a growing eco-industry to economic sustainability', *Energy Policy*, **48**, 13–21, doi 10.1016/j.enpol.2012.04.045.

McKinnon, A.C. (2007), 'Decoupling of road freight transport and economic growth trends in the UK: an exploratory analysis', *Transport Reviews*, **21**(1), 37–64.

McKinsey & Company (2009), *Pathways to a Low-carbon Economy*, McKinsey & Company.

Metz, D. (2010), 'Saturation of demand for daily travel', *Transport Reviews*, **30**(5), 659–74.

Millard-Ball, A. and L. Schipper (2011), 'Are we reaching peak travel? Trends in passenger transport in eight industrialized countries', *Transport Reviews*, **31**(3), 357–78.

Netherlands Environmental Assessment Agency and Stockholm Resilience Centre (2009), *Getting into the Right Lane for 2050 – A Premier for EU Debate*, Bilthoven: Netherlands Environmental Assessment Agency and Stockholm Resilience Centre, October.

Nordbakke, S. and T. Schwanen (forthcoming), 'Wellbeing and mobility: a theoretical framework and literature review focusing on older people', *Mobilities*.

Pooley, C.G. (2010), 'Landscapes without the car: a counterfactual historical geography of twentieth-century Britain', *Journal of Historical Geography*, **36**(3), 266–75.

Schwanen, T., D. Banister and J. Anable (2011), 'Scientific research about climate change mitigation in transport: a critical review', *Transportation Research Part A: Policy and Practice*, **45**(10), 993–1006.

Sen, A. (1980), *Equality of What? Choice, Welfare and Measurement*, Cambridge, MA: Harvard University Press.

Sen, A. (1993), 'Capability and wellbeing', in M. Nussbaum and A. Sen (eds), *The Quality of Life*, Oxford: Oxford University Press, pp. 30–53.

Sen, A. (2009), *The Idea of Justice*, London: Penguin Books.

Skinner, I., H. Van Essen, R. Smokers and N. Hill (2010), *Towards the Decarbonisation of the EU's Transport Sector by 2050*, Project final report, AEA/ED45405/Final Report, London.

Sperling, D. and D. Gordon (2009), *Two Billion Cars – Driving Towards Sustainability*, Oxford: Oxford University Press.

Tapio, P. (2005), 'Towards a theory of decoupling: degrees of decoupling in the EU and the case of road traffic in Finland between 1970 and 2001', *Transport Policy*, **12**(2), 137–51.

Tapio, P., D. Banister, J. Luukkanena, J. Vehmasa and R. Willamoc (2007), 'Energy and transport in comparison: immaterialisation, dematerialisation and decarbonisation in the EU15 between 1970 and 2000', *Energy Policy*, **35**(1), 433–51.

Tapio, P., N. Finel, D. Banister, J. Luukkanen, J. Vehmas and R. Willamo (2011), 'Global decarbonisation patterns – total and transport $CO_2$ intensity', Working Paper no. 1056, Transport Studies Unit, School of Geography and the Environment, Oxford University, October.

Tight, M., P. Timms, D. Banister et al. (2011), 'Visions for a walking and cycling focussed urban transport system', *Journal of Transport Geography*, **19**(6), 1580–9.

Tuominen, A. and T. Ahlqvist (2010), 'Is the transport system becoming ubiquitous? Socio-technical roadmapping as a tool for integrating the development of transport policies and intelligent transport systems and services in Finland', *Technological Forecasting and Social Change*, **77**(1), 120–34.

Unruh, G.C. (2000), 'Understanding carbon lock-in', *Energy Policy*, **28**, 817–30.

Urry, J. (2007), *Mobilities*, Cambridge: Polity Press.

Van den Bergh, J.C.J.M. (2009), 'The GDP paradox', *Journal of Economic Psychology*, **30**(2), 117–35.

Van der Heijden, K. (2005), *Scenarios – The Art of Strategic Conversation*, Chichester: John Wiley & Sons.

Vehmas, J., P. Malaska, J. Luukkanen et al. (2003), 'Europe in the global battle of sustainability: rebound strikes back?', Advanced Sustainability Analysis, Publications of the Turku School of Economics and Business Administration, Series Discussion and Working Papers 7, Turku, Finland.

Wells, P. and P. Nieuwenhuis (2006), 'Business models for relocalisation to deliver sustainability', in P. Nieuwenhuis, P. Vergragt and P. Wells (eds), *The Business of Sustainable Mobility: From Vision to Reality*, Sheffield: Greenleaf Publishing, pp. 92–102.

# 14. Sociotechnical transition in the transport system

**Tim Schwanen**

## INTRODUCTION

While the notion of sustainability has been around – and contested – for several decades, a new consensus seems to be emerging sparked in part by the imminent crises of global climate change and oil shortage. At least in Europe more and more actors in the corporate world, civic society, science and the media as well as policy-makers and politicians, now accept that fundamental, systemic changes in the flows and circulation of people, goods and energy are required. The term 'transitions' is increasingly used to denote such systemic changes and the adjective 'sociotechnical' is often added to foreground their comprehensiveness. Indeed, a sociotechnical transition has been defined as a major shift in which the technologies, markets, infrastructures, consumer practices, cultural meanings, policies and regulations, and scientific knowledges that together constitute a sociotechnical system are durably reconfigured (Geels, 2004).

Couching the shift towards low carbon mobility as a sociotechnical transition is useful for multiple reasons. Beyond emphasizing the need for changes in the basic architecture of contemporary mobility systems, transition thinking draws attention to a wider range of actors than is common in transport research (Schwanen et al., 2011): not only public authorities, urban planners, transport agencies, vehicle manufacturers and consumers are important, so are venture capital suppliers, insurance companies, designers, material and machine suppliers, lobby organizations, social movements, the media and research institutes. Further, sociotechnical transition thinking examines the linkages and interactions between all these actors and the ways in which their practices and competences shape each other and become aligned, de-aligned and realigned. Sociotechnical transition thinking is also evolutionary and process oriented. It is therefore capable of addressing the key philosophical problematic of continuity and change – being and becoming – and draws attention to both path dependencies and habits, and tipping points and the emergence of truly new events (that are unexpected and unforeseen in light of historical developments). As such, it moves beyond the static character, emphasis on predictability and narrow understanding of rationality of neoclassical economics, which pervades transport research and practice so profoundly. Moreover, transition thinking puts innovation centre-stage – a process that has attracted less critical examination in transport studies than one would expect on the basis of the field's considerable interest in technofixes for reducing transport's negative externalities.

Thinking about sociotechnical transitions comes, however, in many varieties. One could mobilize, for instance, evolutionary economic theory (Dosi, 1982; Nelson and Winter, 1982); long-wave economic theory (Freeman and Louçã, 2001); the multi-level perspective and affiliated approaches (Nill and Kemp, 2009; Geels, 2011); practice theory (Shove and Walker, 2010); and sociologies of complexity (Urry, 2011). Nonetheless, while expanding rapidly, applications of transition thinking in transport and mobilities research are as yet fairly limited in number and tend to be animated by the multi-level perspective, practice theory and sociologies of complexity. These perspectives are also broader in terms of underpinning worldview than neo-Schumpeterian economic theory. It is for these reasons that this chapter concentrates on the multi-level and affiliated perspectives, practice theory and sociologies of complexity, providing state-of-the-art summaries of key ideas and drawing out the most important lessons for the shift towards low carbon mobility. Regarding the latter, the emphasis is on the system of automobility built around the internal combustion engine-powered (ICE-powered) car. This is done for the sake of consistency and because this system is the single most important source of carbon emissions from transport.

The chapter concentrates more on the theoretical perspectives and how they help us to understand the social reality of change in (automobile) transport than on the implications for policy-making, which are picked up in the concluding chapter (Chapter 16). This is first of all because the thinking on sociotechnical transitions by innovation scholars, sociologists and geographers differs fundamentally from mainstream perspectives on dynamics in transport systems among transport researchers and urban planners. It is also because systemic change in personal transport insofar as it can be governed at all requires a much broader set of policy initiatives than the usual package of economic incentives; publicity campaigns; facilitating technology improvement, including fuel efficiency and alternative power-trains; infrastructural interventions; and better integration of land use and transport. These measures are all useful and perhaps even necessary for step change but they are unlikely to be sufficient. What exactly that broader set of required interventions should consist of is, however, not immediately clear. As a first step towards tackling this challenge we first need to better understand how step changes come about in personal transport systems.

## MULTI-LEVEL AND AFFILIATED PERSPECTIVES

### Key Problematic

First articulated in the Netherlands in the 1990s (Rip and Kemp, 1998), the multi-level perspective (MLP) brought together evolutionary economics and social constructivist approaches within Science, Technology and Society (STS) approaches, which have since been augmented with insights from sociology and political science. The theoretical approach has been developed in conjunction with historical case studies of past transitions in transport, sanitation, water supply and industrial production, as well as research on contemporary transitions in electricity systems, transport, organic food and sustainable housing (Geels, 2012). Strategic niche management (Kemp et al., 1998), transition management (Rotmans et al., 2001) and time strategies (Nill and Kemp, 2009) are partially based on the MLP. The main difference is that these three approaches are oriented towards facilitating and enacting transitions in practice, whereas the MLP as elaborated by Frank Geels and colleagues seeks to describe and explain sociotechnical transitions. The basic problem all four perspectives consider is nonetheless similar: how does and can systemic change occur given the various lock-in[1] mechanisms that stabilize the existing sociotechnical system?

A wide range of such lock-in mechanisms can be identified (Kemp et al., 1998; Unruh, 2000; Banister et al., 2011):

- Technology and production. Upon invention new technologies need to be optimized, redesigned and tested on a large scale by consumers before they can become competitive and economies of scale can be created. Incumbent firms are often hesitant to invest in radical inventions because they may render core competencies and existing production facilities superfluous, demand large sunk investments in new facilities and may not appeal to potential consumers. These factors are particularly important in sectors where the market is dominated by a few large firms, as with the auto industry. Banks and other capital suppliers are often risk aversive as well.
- Infrastructure. New technologies often require adaptations of existing infrastructures or the development of new infrastructures but

this requires large, risky investments that go against the grain of the sunk investments of existing infrastructures.

- Consumer practices and lifestyles. These have often become attuned to particular technologies, including the ICE-powered car. Consumers may therefore resist systemic change. Corporate actors and banks often anticipate such resistance, which tends to increase their risk aversion.
- Cultural meanings. Values become attached to technological artefacts and associated practices, which then obtain symbolic connotations and become part of identity construction. The connotations of car use with freedom, flexibility and responsible parenting (Dowling, 2000; Freudental-Pedersen, 2009) exemplify these processes well.
- Regulation, policy-making and technoscientific knowledge. Institutional commitments to the existing system are often strong and incumbent actors are often effective and powerful in lobbying policy-makers and politicians. And when the policy rhetoric changes, it usually does so more quickly than practice, which is often haunted by compartmentalized responsibilities, reliant on instruments that favour incremental over radical change, and dependent on research and expertise that reflects the built-in logics of the existing system. Additionally, new generations of transport professionals are also 'acculturated' into status quo forms of knowledge and thinking through training and education. Financial arrangements and tax regimes matter greatly as well: the absence of adequate pricing of the negative externalities of carbon-intensive forms of transport is deeply entrenched in most countries, which helps to perpetuate the system of automobility.

## The MLP's Central Ideas

The above discussion suggests that a sociotechnical system has a strong tendency towards reproduction and incremental change, and transition scholars use the term 'regime' in this context (Rip and Kemp, 1998). A sociotechnical regime is the 'semi-coherent set of rules that orient and coordinate the activities of the social groups that reproduce the various elements of socio-technical systems' (Geels, 2011, p. 27). Examples of such rules are cognitive routines and shared beliefs, competences, lifestyles, institutions, regulations and contracts. The technology, markets, user practices, culture, politics and technoscientific knowledge that make up a particular regime all have their own rules, but these are interconnected and change in one often, but not always, leads to change in the others (Geels, 2004), hence the term 'semi-coherent' in the above definition of a regime.

In the MLP the sociotechnical regime constitutes the meso level; it is complemented by the macro level of the sociotechnical landscape and the micro level of niches (Rip and Kemp, 1998). Niches are protected spaces – typically market niches, demonstration projects or research and development (R&D) laboratories – where niche actors experiment with emerging innovations that differ from the regime and where users have specific demands. The landscape, in contrast, is the exterior context that affects regimes and niches and over which actors in a regime or a niche have little control. It can be considered given and comprises, for instance, demographic developments like population ageing, macroeconomic conditions and anthropogenic climate change. Niche, regime and landscape have often been conceptualized as a nested hierarchy but Geels (2011) distanced himself from this metaphor, arguing that the different levels refer first and foremost to the degree of stability or structuration of practices (after Giddens, 1984). Nonetheless, others (Van Driel and Schot, 2005) have highlighted that landscapes exhibit differentiated dynamics: some factors change only on very long timescales (for example, biological evolution), others change over the course of decades (for example, population ageing, globalization), still others through rapid shocks (for example, oil prices).

The MLP's central idea is that systemic change emerges from the interactions between niche, regime and landscape dynamics. These interactions can take many different forms but one pattern has been privileged in early publications (Geels, 2002). It is the most well known and can be considered the 'classic' pattern; it comprises three interrelated processes (Figure 14.1):

- Niche innovations build up internal momentum over time.
- Landscape changes put the regime under pressure.
- Regime destabilization offers windows of opportunity for niche innovations to be scaled up, displace the old and establish a new regime.

Of these processes, the first has been studied most extensively by transition scholars. For Kemp et al. (1998), niche innovations gain momentum through the building of social networks around a new technology in which more and more diverse actors become enrolled, and through collective learning processes. Through practical experiments – for instance, the introduction of 350 lightweight electric vehicles for everyday use in Mendrisio, Switzerland in the 1990s (Hoogma et al., 2002) – niche actors learn about technical design, production, infrastructure, markets, cultural meaning and regulation and policy-making. Rotmans et al. (2001) also highlight the importance of longer-term visioning in such experiments. More recent work has sought to understand how networks of niche actors

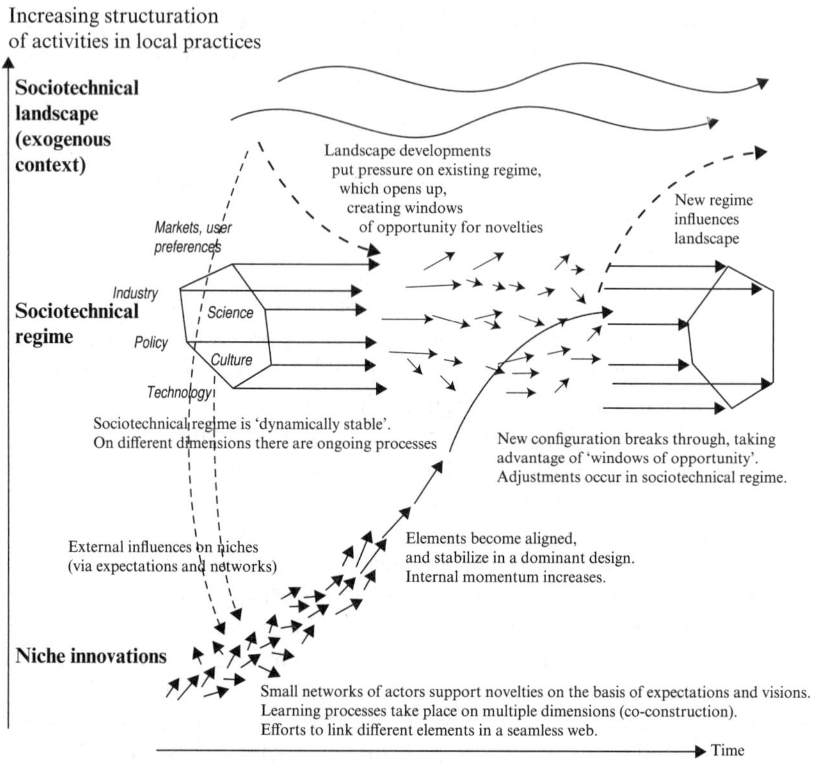


*Source:* Geels (2012).

*Figure 14.1 The 'classic' interaction pattern*

develop and learn over the course of multiple experiments and projects. To this end, Raven and Geels (2010) distinguish two sublevels within niches that are dialectically related with each other:

- The local level of practices and individual projects.
- The global niche level of the emergent community and shared cognitive rules, such as technical models, heuristics and beliefs.

The shared cognitive rules at the global niche level are created through aggregation of experiences in local projects and practices. At the same time, those rules function as resources for the articulation of expectations, which guide the formation of new local projects. In collective processes within the emergent community the outcomes of those projects are interpreted and lessons are drawn through processes of aggregation and codification. This reinforces or modifies the shared cognitive rules. Positive outcomes facilitate the enrolment of additional actors and expansion of the resource base; negative outcomes may diminish social networks and available resources, lower expectations and even trigger a niche's demise.

In recent years attention for other than the 'classic' multi-level interaction pattern has increased. Geels and Schot (2007) proposed a wider range of multi-level interactions, which vary in terms of the timing of landscape vis-à-vis niche developments and the degree to which those developments have disruptive or symbiotic relationships with regime dynamics:

- Transformation. In this type moderately disruptive landscape pressures in conjunction with premature niche developments tend to result in regime actors modifying the direction of development trajectories whist leaving the system's basic structure intact. Cumulative adjustment and reorientation of regime rules is the result: the newer regime gradually grows out of the older but is not fundamentally different. This pathway is exemplified by the transition from cesspools to sewer systems in the Netherlands in the late nineteenth and early twentieth centuries (see Geels, 2006a).
- De-alignment and realignment. Strong and sudden landscape change triggers regime disintegration and the absence of a mature niche implies a time of uncertainty in which multiple niches compete. Stabilization eventually takes place around one innovation, constituting the basis for a new regime. The transition from horse-drawn carriages to cars around 1900 exemplifies this path. At that time major changes, like urbanization, immigration, a growing middle class and improved hygiene created rapidly accumulating problems for horse-based travel. The niches of electric trams, bikes, electric cars and ICE-powered cars competed for several decades but eventually the technically inferior petrol-fuelled car displaced the other technologies, which at the end of the nineteenth century were more fuel efficient and whose performance was more reliable. A series of more-or-less accidental events ranging from the fact that a petrol-fuelled vehicle was one of only two to complete a horseless carriage competition in Chicago in 1896 and Ford's invention of processes to mass produce cars led to the development of the system of automobility (Urry, 2007).
- Technological substitution (effectively the pattern labelled 'classic' above). A combination of strong and sudden landscape pressures and the existence of a mature niche lead to regime displacement. Geels (2002) describes the shift from sailing to steamship in mid-nineteenth-century Britain as an example of this pathway. Critical

landscape developments were mass emigration from Europe to America in the wake of the 1848 revolutions and the potato famine in Ireland, as well as the opening of the Suez Canal. They boosted economic competition in which steamships, whose design and performance had matured over the first half of the nineteenth century, displaced sailing ships.

- Reconfiguration. Symbiotic niche developments are incorporated by regime actors to solve local problems but subsequently and gradually trigger more fundamental changes to the regime's basic architecture. The transition from traditional factories to mass production in the US in the nineteenth and early twentieth centuries exemplifies this pathway. A sequences of new elements and procedures, such as machine tools, division of labour and assembly, became incorporated in factories, which gradually led to fundamentally different systems of production, particularly in the automobile industry (Geels, 2006b).

While useful, this typology by no means exhausts the possibilities (Geels, 2011). Further differentiations are possible on the basis of the relative power of landscape and bottom-up changes, whether change is driven by existing elites or social movements on the outside, and the extent to which changes are carried by broader collectives of actors or driven by the visionary leadership of particular individuals.

## Governing Multi-level Interactions in Practice

The approaches of strategic niche management (SNM), transition management (TM) and time strategies (TS) essentially work with the MLP's logic in, and extend it to, practice. SNM focused on strategies through which niche developments can be strengthened and drew specific suggestions for policy-makers. Kemp et al. (1998) and others have stated clearly that price incentives have a role to play in transitions but are insufficient to trigger them; policy-makers should act as enablers and catalysts rather than regulators or technology sponsors. As a governance approach SNM has more or less been displaced by TM, which has a broader scope, focuses on system change and has been described as 'meta-governance' (Loorbach and Rotmans, 2010). Key elements of TM are the utilization of ongoing trends for societal goals (goal modulation); a focus on long-term, strategic thinking for short-term action; consideration for linkages across such domains as transport, energy and agriculture; and an emphasis on reflexive learning about multiple transition paths to avoid early lock-ins (Nill and Kemp, 2009; Loorbach, 2010). A central role is played by the devel-

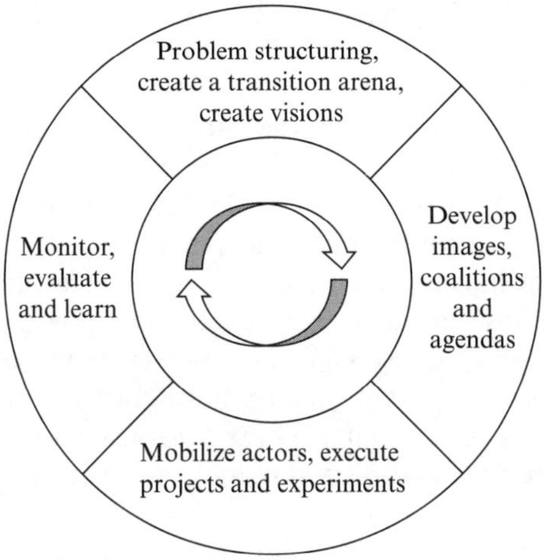


*Source:* After Loorbach (2010).

*Figure 14.2 The transition management cycle*

opment of joint visions in the 'transition arena' in which multiple actors are brought together to align their actions and make the collective more competitive in the marketplace (Figure 14.2). TM has been employed in a wide variety of contexts, including the healthcare, energy and transport sectors in the Netherlands (details in Loorbach and Rotmans, 2010; Kemp et al., 2011).

TS is arguably the least developed approach. Its starting point is that the degree of a system's lock-in varies over time: tensions within a regime, landscape pressures or the development of strong niches can create greater or smaller windows of opportunity for change. The idea is that different types of policies are needed over time, depending on the strength of the regime, maturity of niches and landscape pressure (Nill and Kemp, 2009):

- Window preparation. When the regime is strong, policies are needed that stimulate a diversity of alternatives and allow niches to build up momentum.
- Window creation. When both the regime and landscape pressures are strong, policies have to avoid quick fixes (add-on technologies or retrofitting) trumping more radical change.
- Window utilization. When the regime is weak, policies can seek to enact systemic change by stimulating or favouring one niche innovation.

## THE MLP AND PASSENGER TRANSPORT

Various studies have applied the MLP, SNM and TM to examine and speed up systemic change towards low carbon transport (see, for instance, Hoogma et al., 2002; Nykvist and Whitmarsh, 2008; Dijk and Yarime, 2010; Van Bree et al., 2010; Avelino, 2011; Kemp et al., 2011; Geels, 2012; Geels et al., 2012). Unsurprisingly, the 'regime' concept is coupled to the dominance of the ICE-powered car in user practices, planning, cultural expressions, technoscientific knowledge and the power of car manufacturers and lobby groups. Geels (2012), however, points out that there are subaltern regimes organized around train/rail, bus and cycling beyond the dominant regime of automobility. Those other transport modes have been around for a long time and their systems are too stable to be considered niches.

Equally unsurprising is that much research has considered the identification and evolution of existing niches. There is a bias towards high-tech solutions – in particular electric vehicles – but at least two studies have identified a broader range of niches. For the UK, Nykvist and Whitmarsh (2008) have enumerated three important niches:

- Radical vehicle technologies, such as battery electric vehicles (BEVs) and hybrid electric vehicles (HEVs) and alternative power-train technologies.
- Product-to-service shift through carpooling/sharing and more public transport.
- Mobility management focusing on reduced travel through stimulation of cycling and walking, information and communications technologies (ICTs) and pricing measures.

Synthesizing recent development in the UK and the Netherlands, Geels (2012; see also Geels et al., 2012) has produced a similar if more elaborate list:

- Intermodal travel: initiatives to seamlessly weave together rail, bus, taxi, car and walking/cycling, exemplified among others by park-and-ride schemes.
- Cultural and sociospatial niches: these challenge the normality of automobility through land use (transit-oriented development, compact city) or collective ownership as with car/bike-sharing programmes.
- Demand management through personalized travel planning, information, awareness campaigns and so on.

- Public transport innovations primarily through new infrastructure, better information provision and modernized vehicles.
- ICTs for increasing transport's efficiency (intelligent transport systems) and substituting physical movement (telecommuting, e-commerce).
- Green propulsion technologies, such as BEVs and HEVs.

For Geels, most of these niches have only limited internal momentum and the technology-driven niches of green propulsion and ICTs have the most potential. Others (Nykvist and Whitmarsh, 2008; Kemp et al., 2011) seem to underwrite the relative strength of the green propulsion niche. Significant barriers for this niche can nonetheless be identified. For one, the market for ICE-powered cars is so segmented that it is more appropriate to speak of different (sub)markets where the functional attributes of and cultural meanings inscribed into cars often differ starkly. Thus, people interested in buying, say, sport utility vehicles (SUVs) will generally not consider a city car as an alternative. The implication is that green propulsion niches have to become competitive not in a single market but in many submarkets and thus be able to meet a wide range of consumer expectations and cultural values (Dijk, 2011).

Another barrier is that there exist multiple green propulsion technologies that each have their own hype-disappointment cycles (Geels, 2012). At present, BEVs are on the rise again after an earlier hype in the early 1990s and a prolonged period of disappointment and limited interest between 1995 and 2005. It is as yet too early to write off hydrogen and biofuel vehicles and to identify a 'winner'. The uncertainty this creates has so far dampened incumbent manufacturers' enthusiasm. Indeed, the attempts of environmental regulators to stimulate radical innovations in car engines since 1990 have been (mis)used by car manufacturers to stimulate incremental change in ICEs. Policy stimuli have effectively entrenched the current system's lock-in (Dijk and Yarime, 2010). The barriers that Geels and Dijk outline are amplified by the particularities of the automotive industry: not only are the initial capital investments extremely large, there are also very high barriers to the entry of new competitors, which means that incumbent manufacturers almost certainly play a decisive role in the propulsion niche's fate (Zapata and Nieuwenhuis, 2010).

The automobility regime has become somewhat weaker over the past two decades, especially in (the central parts of) cities and among young adults, some of whom do not obtain their driver's licence at the same age as earlier generations did (Nykvist and Whitmarsh, 2008; Geels, 2012). On the whole, however, the automobility regime is remarkably resilient due to various causes. The landscape pressures of climate change, oil scarcity and the emergence of the information society are outweighed by other landscape tendencies that stabilize the regime (Geels, 2012; Geels et al., 2012). The latter include: cultural preferences for private rather than collective ownership; the cultural valorization of speed and time efficiency; prevailing understandings of social success; individualization; and growing markets for ICE-powered cars in emerging economies. Sustainability and climate change are also relatively minor concerns for regime actors, and when transport policy-makers are genuinely committed to emission reduction, they are constrained by dependencies on the wider public and the car industry, which have become stronger with deregulation and devolution of policy-making under neoliberalism and the growing role of the popular media in scrutinizing government and in 'making and breaking' the reputation of businesses (see also Urry, 2011).

In terms of transition trajectory, both Nykvist and Whitmarsh (2008) and Geels (2012) suggest that the 'reconfiguration' path is the most likely, at least in Western countries with mature automobile regimes. According to Geels, the car-based transport system will probably remain dominant in the coming decades, although a gradual greening of technology will probably occur. The niches of intermodal transport, cultural and sociospatial innovations and demand management will for the most part remain just that – niches – even though they may to some extent be incorporated into the regime insofar as they can solve (locally specific) problems in the latter and do not challenge its basic structure. Either way, intermodal transport and demand management will be largely concentrated in cities.

Drawing lessons from past sociotechnical transitions for a systemic shift towards low carbon mobility is fraught with problems. Nonetheless, historic research shows that it takes at least three decades to complete transitions built around technologies and associated infrastructure and led by (newly emerged) businesses, policy-makers and other institutional actors (Geels, 2012). For passenger transport in Western countries, this timespan is squarely at odds with the urgent need for decarbonization in light of the growing oil consumption in transport in emerging economies and in international transport and trade (Banister et al., 2011). Passenger transport is likely to remain a key source of carbon consumption for some time.

## BEYOND THE MLP

### Sympathetic Critiques

While useful, the MLP comes with a number of inevitable blind spots. As with any analytical perspective used to make sense of reality, it works like a pair of spectacles: it brings certain parts of reality into sharp focus but blurs other parts. Given that we as researchers do not have direct (visual) access to reality, doing away with visual aids entirely is not possible. Putting on other aids like sunglasses or contact lenses is therefore important and raises awareness about blind spots that would otherwise go unnoticed. Putting on those aids is exactly what researchers from a range of disciplinary backgrounds have done, and which has resulted in various criticisms of the MLP, SNM and TM (see, for instance, Shove and Walker, 2007; Genus and Coles, 2008; Voß et al., 2009; Lawhon and Murphy, 2012). Transition scholars have been keen to address criticisms. Avelino (2011), for instance, has addressed the lack of attention for politics and power struggles between stakeholders in much MLP and TM literature to which various critics have pointed.

A number of issues nonetheless remain. First, transition thinking tends to privilege the dynamics in technology over those in other regime components (Lawhon and Murphy, 2012; Shove, 2012). Systemic change is often narrowed down to technology-centred change whereby technical artefacts, such as electric vehicles or rental bikes, and associated infrastructures are highlighted over user practices and cultural meanings. The underlying assumption tends to be that environmentally more benign forms of supply will, once in place, displace carbon-intensive forms and that this will eventually induce change in other regime components. Recognizing the relative neglect of culture, Geels and Verhees (2011) propose to analyse the cultural acceptance of new technologies in terms of processes of contestation in which different social groups struggle over interpretations of specific technologies (frames) and more general ways of thinking about technology (discourses). Studying vehicle technologies, Dijk (2011) has also examined changes over time in frames, understood as the ways in which innovations are described and interpreted by actors.

Yet, the work by Geels, Dijk and others also illustrates other shortcomings of the MLP. Second, then, culture is reduced to meaning and cognition, which means that a logocentric[2] conception of agency – I think, therefore I act – is ascribed to the general public. This contravenes the fact that most actions of (potential) users of a transport system are non-reflective, extra-discursive and habitual. Car drivers do not make conscious tradeoffs between different modes or think about how driving imbues them with freedom, control or autonomy each time they get onto the road. As centuries of philosophy, decades of social theory and years of neuroscience have shown, the relation between what people do and what they think or put into words is much more complex than logocentric understandings of agency and culture suggest (see, for instance, Schwanen et al., 2012). Related to this, and third, the subjectivities of the general public – the ways of being and doing ascribed to its members – are overly restricted in MLP research: they are considered either parts of social movements (as in Geels's recent work), or understood in binary terms as adopters and non-adopters. The consequence of the narrow interpretations of culture and the public is that not much thought has been given to innovation trajectories whereby the general public rather than businesses and political elites are taking the lead in innovation processes from very early onwards and in ways whereby competition in the marketplace is not necessarily involved.

A fourth concern is that critical questions about social equity and justice in relation to innovations – Who benefits, when, where and how? Who loses? – are not normally asked in MLP studies. A fifth is that the MLP and related approaches suffer from naivety regarding space and place in various ways. Insofar as geographies are considered, these are assumed to be organized on the basis of a territorial logic of vertical relations with regimes at the national level and niches occurring in specific locations – often cities – within these. The fact that space and place increasingly embody network logics in which horizontal relations are dominant is not fully considered: the fate of niches and regimes is not simply a consequence of processes in a particular country or city, but of networked developments in numerous other sites across a wide range of spatial scales as well. Hodson and Marvin (2009), for instance, show how London's hydrogen economy initiatives have emerged out of relations between agents and processes across scales from the neighbourhood to the levels of the European Union (EU) and transnational companies.

At the same time, MLP studies do not properly recognize the 'spatial embeddedness' of both niches and regimes in locally clustered networks of firms and other stakeholders. Economic geographers have long been at pains to point out that innovations do not thrive equally in different places (see also Truffer and Coenen, 2012): institutional thickness – the totality of social, cultural and institutional support available to businesses – varies geographically, and so may the success of niche innovations. From a geography perspective, then, the tidy separation of niche, regime and landscape is incapable of representing the geographical complexity of, and hence the power relations inherent to, innovation processes. At best naive and at worst faulty recommendations regarding transition governance may result. Further, MLP-informed empirical research draws disproportionally on evidence from Northwest Europe – the Netherlands in particular – which has significant implications (Banister et al., 2011; Lawhon and Murphy, 2012): insights and policy prescriptions about transitions derived there may not hold elsewhere, particularly in the Global South where government is often weaker and countries are positioned differently in global commodity chains.

Given these concerns, it is not surprising that sociologists and geographers have extended and complemented the MLP in various ways. Two strands of work are particularly useful in addressing some of the above criticisms and helping us to better understand the potential of a systemic shift towards low carbon transport: the version of practice theory developed by Elizabeth Shove and colleagues and John Urry's work on the sociologies of complexity.

**Theories of Social Practices**

Practice theory is a tradition in sociology, human geography and related disciplines that argues that societies, systems, social processes and technological change are produced in and through 'practices'. A practice is here understood as a routinized nexus of sayings and doings (Schatzki, 2009) and the integration of elements from multiple realms in the 'doing' of human activity (Reckwitz, 2002; Schatzki, 2009):

- The material – human bodies, objects, equipment, infrastructures.
- The procedural – rules, laws, know-how, competence, skill.
- The cognitive – thoughts, ideas.
- The symbolic – meanings, discourses, norms, identities.
- The affective – feelings, emotions, atmospheres.

Driving, for instance, is a practice in which all these elements are continually interwoven: a car, roads, driving skills, notions of speed, efficiency and freedom, road rage, traffic rules and so forth. A transition of certain type of practices amounts to one or more of the following (Watson, 2012):

- The elements comprising the practice change.
- The population of 'carriers' – the people performing them – changes.
- Other practices coupled to the practice of interest change (Watson, 2012).

The coevolution of practices is well exemplified by the increased car use in Western countries for driving children around over past decades. This is linked to children's increased participation in extra-curricular activities and structured leisure; the rise in mothers' labour force participation; greater conflicts between the activity schedules of household members; suburbanization and urban sprawl; greater concerns over children's (social) safety and fear of crime; and car trips as opportunities for being together, intimate conversations and fun (Dowling, 2000; Schwanen, 2011).

Thinking of practices as interweaving elements highlights the arbitrariness of privileging technological artefacts and infrastructures that also characterizes MLP studies of sustainable transport and TM experiments in (Dutch) policy practice: transitions can also involve other elements implicated in practices. In her research on energy use within the home Shove (2012) focuses on how the cultural notion of maintaining the temperature at 21–22 °C rather than air conditioning has moved from innovation to normality in different geographical settings. Her work suggests, then, that analyses of the potentialities of widespread low carbon mobility should not be limited to technology, infrastructure and policymaking. Attention is also warranted for whether the routines of walking, cycling, using public transport or even a BEV are compatible with the complex rhythms of everyday life in dual-earner families and other contemporary household types, and for whether low carbon transport modes are as reliable as allies in the construction of successful do-it-yourself biographies as ICE-powered cars (Freudendal-Pedersen, 2009). Multiple trajectories of change need to be analysed in conjunction for understanding sociotechnical transitions.

For Shove (2012), those trajectories can be analysed without the conceptual baggage of niche and landscape. Transitions can also be understood as a reconfiguration over time in the interconnections between the technology, infrastructure, markets, user practices, culture, policies and technoscientific knowledges that make up a regime. This is possible because all these are enacted in, or are themselves, practices; sociotechnical systems are systems of practices (Watson, 2012). Thinking along these lines has multiple advantages. For one, human agency – particularly non-logocentric forms in which conscious thought is not the driving force – is foregrounded to a greater extent than in the MLP. Practices are routinized and habitual but outcomes are never given in advance: there is always an element of creativity and possibility of doing things differently than on previous occasions, and this may be the root for larger-scale change (see also Law, 2004).

Second, practice theory helps us to better understand how culture changes. For culture is neither an exogenous landscape development beyond regime and niche actors' control nor a series of meanings discursively articulated as frames or discourses. It is a multiplicity of interwoven doings from which meanings proceed. Third, focusing on regimes as bundles of practices allows for more straightforward analysis than with the MLP of the circulation of the elements implicated in practices across space through networks of horizontal relations; the multi-scalar geographies of transitions can be explored more satisfactorily than with the MLP. We can thus begin to think about why it is that the cultural convention among middles-class households to weave together home, work, shopping and leisure with cars is now also becoming the norm in Africa's emerging economies, including Ghana (*Guardian*, 2011) with all the implications this will have for reducing the global growth in fossil fuel usage in transport. Finally, practice theory emphasizes that policy-making is itself a set of practices shaped by others. It reinforces the MLP view that policy-makers are not 'outside' the system and can only intervene on the basis of what previously enacted practices have made possible, acceptable and sensible: they have only limited room for triggering change. At the same time, practice theorists also contend that there is always a potential for change in what policy-makers, and other agents, do: '[e]very action . . . is the restarting of history; every action is potentially the beginning of something new' (Schatzki, 2009, p. 216).

Applications of practice theories to understand transitions towards sustainable mobility have been rare to date but Shove and Walker's (2010) analysis of congestion charging in London showcases their potential. These authors argue that the scheme is successful not simply because policy-makers were able to implement it. More important have been the cumulative responses of Londoners who have integrated the change into their highly diversified rhythms of everyday life. This outcome was never given in advance (notwithstanding the predictions mainstream transport models generated); it was – and is – unstable, dynamic and reversible. Effective innovation is more than recruiting adopters and scale jumping; it is about extending the repertoires of elements that can be integrated into mobility practices and everyday life more generally.

**Sociologies of Complexity**

For almost a decade John Urry has promoted a complexity approach to mobility systems (Urry, 2004, 2007, 2011). Suffice it to say that his conceptualization of the locked-in system of automobility bears clear similarities to the lock-in mechanisms and the MLP's regime concept discussed above; here the recent work of Urry (2011) on innovation is highlighted. Urry conceptualizes innovation as a complex process of combination and synchronization. It typically consists of existing elements of machinery, text, technology and organizations becoming combined in new configurations and in the practices of many different actors that stretch across local, national and global levels becoming synchronized. Synchronization tends to occur around some idea, artefact or system that then becomes fashionable. It is well exemplified by the rapid diffusion and uptake of the mobile phone in the 2000s in Western countries and with the spread of Occupy camps in 2011.

Urry's perspective complements the MLP and practice theories in various ways. It challenges the preoccupation with high-tech solutions for perceived problems by foregrounding the importance of 'existing' elements, which are reshaped through enrolment in new relations and configurations. Urry and colleagues (Tyfield et al., 2010) emphasize disruptive innovation, whereby cheaper, easier to use alternatives to existing products or services are produced by non-incumbents. It is primarily characterized by a social redefinition of a technology and by the likely offering in the first instance of lower than cutting-edge functionality for different uses and to neglected customers. One example is the conversion package the Dutch company e-Bike-Kit.nl offers for turning a standard bike into an electric bike (http://www.e-bike-kit.nl).

E-bike-Kit.nl highlights another crucial feature of Urry's (2011) perspective: the importance of 'tinkering' with new technologies, which are gradually improved by (feedback from) enthusiastic practitioners. Urry (2011, p. 131) discusses how Bertha Benz's tinkering played a seminal role in turning the car into a fashionable machine for living. 'Users', then, are not passive (non-)adopters. In many ways they are the innovators who play with, and integrate, elements into daily routines. This is also shown by Lyons and colleagues (2012) who list a number of user innovations in transport, some of which could be classified as disruptive innovations. Many of these rely on the affordances of the internet (http://www.mybikelane.com, for instance) but not all. The latter are exemplified by 'slugging', which is a form of ad hoc informal carpooling on commutes.

For passenger transport Urry (2011) nonetheless highlights the importance of electric vehicles as a key carrier of a low carbon mobility system. This apparent contradiction with the previous paragraphs dissolves if we consider two things. Public transport and cycling/walking cannot offer similar levels of flexibility, comfort and security as cars, and a successor to the ICE-powered car will have to offer at least similar levels of flexibility if it is to become the standard. Additionally, a transition towards electric vehicles is, for Urry, closely entwined with a normalization of urban compaction, carbon budgeting and digital surveillance. This raises difficult ethical questions about the use of such technologies as closed-circuit television (CCTV), data mining software, digital databases and radio frequency identification (RFID) implants for the tracing of vehicles and individuals and about whether mobility is a freedom and a right.

Whatever the nature of innovations, Urry (2011) emphasizes the importance of tipping points to their success. Their diffusion is characterized by ruptures rather than smooth process: at some moment an innovation 'goes viral'. He also contends that the continuing integration of the physical and digital realms is important here (see also Lyons et al., 2012): social media – the successors of Facebook and Twitter? – may well be crucial to promoting low carbon mobility innovations in the near future.

## FINAL REMARKS

The evolutionary approaches of the MLP and TM offer compelling insights into the potentialities of a systemic shift towards low carbon mobility, in particular through their emphasis on process, interacting lock-in mechanisms and niche developments in ways that highlight the role of markets without falling prey to simplistic market ideologies. Nevertheless, conceptual problems remain regarding the role of culture, the agency and subjectivities of 'users', and the geographies of innovation. Shove et al.'s theory of social practices and Urry et al.'s sociologies of complexity go some way towards addressing those concerns but considerable further work is needed. The practice theory approach, for instance, needs to be broadened in focus: practices other than consumption, like those of corporate actors and policy-makers, need to be addressed and the role of markets needs to be theorized more explicitly. There are also issues that none of the theoretical approaches have taken up satisfactorily. Concerns over the social equity and justice of transitions and their democratic legitimacy need urgent attention in conceptualizations of sociotechnical transitions.

These concerns notwithstanding, the emerging scholarship on sociotechnical transitions is valuable in at least three ways from the perspective of how system transitions in transport can be engendered in practice. First, this body of work enables us to think in new ways about business-as-usual prescriptions for transport policy (for example, economic incentives, publicity campaigns, building new infrastructure and technology improvement). Concepts like the sociotechnical regime highlight the interdependence of infrastructure, technology and regulation with technoscientific knowledge and cultural meanings. This means, among other things, that interventions in existing systems can and probably should target a broader range of phenomena and actors: to induce systemic change it is also important to target the sites where meanings and new knowledges are created. To give just two examples: creating a TV programme as popular as *Top Gear* but then celebrating low carbon and non-private car-based transport might be useful, as would stimulating the development of social science understandings of transport and mobility. And even then it is the multiplicative effect of a wide range of interventions that might in some way, somewhere, at some time trigger systemic change.[3] Trying to trigger changes in multiple societal systems simultaneously and placing transport within a wider set of systems – retailing, education, labour markets, social care provision and so on – is equally important. Systemic change in transport cannot be generated from within the transport sector alone.

Second, sociotechnical transition thinking challenges the belief that transitions can be delivered through conventional policy-making mechanisms and existing governmental institutions. Experiences with TM in the transport context in the Netherlands have so far been mixed: it is difficult for TM to become rooted in transport policy and governance (Kemp et al., 2011). This reflects important differences between the governance philosophy at the heart of TM and conventional transport policy-making; the latter is much more top-down, technocratic, driven by technical expertise and modelling, and exclusionary in the sense of involving and listening to fewer stakeholders than TM (Banister et al., 2011; Kemp et al., 2011). But this is not entirely surprising: a critical insight offered by sociotechnical transition thinking is that lock-in mechanisms are powerful and can obstruct transitions towards low carbon, socially equitable passenger transport systems. This is even more so when the lock-in mechanisms are amplified by financial austerity in the Global North; the lack of political leadership needed for reaching global agreements and for turning the promises in national-level laws and agreements into real action in many countries; and the politicization of climate change research and increasing denial of anthropogenic climate change (see also Urry, 2011). The observations on incompatibilities between TM and existing governmental institutions in the few transport-related practical experiments do not invalidate sociotechnical transition thinking as such. Rather they enhance doubts as to whether transitions will be triggered from within existing governmental institutions: policy-makers do not stand 'outside' the systems they seek to change but are entangled with and dependent on numerous stakeholders within existing regimes (Geels, 2012; Watson, 2012).

Third, sociotechnical transition research also sensitizes us towards the possibility of systemic change in passenger transport following two quite different sets of processes. Sudden, more or less unanticipated major shocks to existing systems like the flooding of a global city or war in the Middle East may create a greater sense of urgency, political will and profound changes to governmental institutions and practices at city-regional, national and global levels. While such catastrophes should of course be avoided, their (imminent) occurrence may open up unanticipated and sudden windows of opportunity for radical change. Additionally, the tinkering practices by non-incumbent 'consumer-innovators' may lead to as yet unimaginable reconfigurations of transport technologies and practices that will become fashionable and diffuse through markets, social media and other mechanisms. One important contribution that sociotechnical transition scholars can make, then, is to develop and provide frameworks allowing transport professionals to better understand the opportunities for systemic change these two sets of processes will offer. This can then stimulate new and creative thinking about how interventions by policy-makers and others can maximize the capacities and competencies for engendering systemic change when a greater sense of urgency arises or consumer-innovators come up with promising inventions in (low-end) technologies and social practices.

## NOTES

1. Lock-in is a state in which a system is 'stuck' and which is difficult to change.
2. Logocentrism is a concept originally coined by German philosopher Ludwig Klages (1872–1956) and here used to denote the tendency in Western philosophy and science to privilege the logos, the word or speech act, as central to how people know and act in the world they inhabit.
3. A complicated phrase is used to foreground contingency: insofar as a transition occurs it should be considered a non-deterministic, 'non-linear' outcome of the multiplicity of interventions in particular times and places. Transferability of outcomes across space and time is by no means guaranteed.

## REFERENCES

Avelino, F. (2011), *Power in Transition: Empowering Discourses on Sustainability Transitions*, Rotterdam: Erasmus University Rotterdam.

Banister, D., K. Anderton, D. Bonilla, M. Givoni and T. Schwanen (2011), 'Transportation and the environment', *Annual Review of Environment and Resources*, **36**, 247–70.

Dijk, M. (2011), 'Technological frames of car engines', *Technology in Society*, **33**, 165–80.

Dijk, M. and M. Yarime (2010), 'The emergence of hybrid-electric cars: innovation path creation through co-evolution of supply and demand', *Technological Forecasting and Social Change*, **77**, 1371–90.

Dosi, G. (1982), 'Technological paradigms and technological trajectories: a suggested interpretation of the determinants and directions of technical change', *Research Policy*, **11**, 147–62.

Dowling, R. (2000), 'Cultures of mothering and car use in suburban Sydney: a preliminary analysis', *Geoforum*, **31**, 345–53.

Freeman, C. and F. Louçã (2001), *As Time Goes By: From the Industrial Revolutions to the Information Revolution*, Oxford: Oxford University Press.

Freudendal-Pedersen, M. (2009), *Mobility in Daily Life: Between Freedom and Unfreedom*, Aldershot: Ashgate.

Geels, F.W. (2002), 'Technological transitions as evolutionary reconfiguration processes: a multi-level perspective and a case-study', *Research Policy*, **31**, 1257–74.

Geels, F.W. (2004), 'From sectoral systems of innovation to socio-technical systems: insights about dynamics and change from sociology and institutional theory', *Research Policy*, **33**, 897–920.
Geels, F.W. (2006a), 'The hygienic transition from cesspools to sewer systems (1840–1930): the dynamics of regime transformation', *Research Policy*, **35**, 1069–82.
Geels, F.W. (2006b), 'Major system change through stepwise reconfiguration: a multi-level analysis of the transformation of American factory production (1850–1930)', *Technology in Society*, **28**, 445–76.
Geels, F.W. (2011), 'The multi-level perspective on sustainability transitions: responses to seven criticisms', *Environmental Innovation and Societal Transitions*, **1**, 24–40.
Geels, F.W. (2012), 'A socio-technical analysis of low-carbon transitions: introducing the multi-level perspective into transport studies', *Journal of Transport Geography*, **24**, 471–82.
Geels, F.W. and J. Schot (2007), 'Typology of sociotechnical transition pathways', *Research Policy*, **36**, 399–417.
Geels, F.W. and B. Verhees (2011), 'Cultural legitimacy and framing struggles in innovation journeys: a cultural-performative perspective and a case study of Dutch nuclear energy (1945–1986), *Technological Forecasting and Social Change*, **78**, 910–30.
Geels, F.W., R. Kemp, G. Dudley and G. Lyons (eds) (2012), *Automobility in Transition? A Socio-technical Analysis of Sustainable Transport*, New York: Routledge.
Genus, A. and A.-M. Coles (2008), 'Rethinking the multi-level perspective of technological transitions', *Research Policy*, **37**, 1436–45.
Giddens, A. (1984), *The Constitution of Society: Outline of the Theory of Structuration*, Berkeley, CA: University of California Press.
*Guardian, The* (2011), 'Africa's burgeoning middle class brings hope to a continent', 25 December, available at http://www.guardian.co.uk/world/2011/dec/25/africas-middle-class-hope-continent (accessed 2 April 2012).
Hodson, M. and S. Marvin (2009), 'Cities mediating technological transitions: understanding vision, intermediation and consequences', *Technological Analysis and Strategic Management*, **21**, 515–34.
Hoogma, R., R. Kemp, J. Schot and B. Truffer (2002), *Experimenting for Sustainable Transport: The Approach of Sustainable Niche Development*, London: Spon Press.
Kemp, R., J. Schot and R. Hoogma (1998), 'Regime shifts to sustainability through processes of niche formation: the approach of strategic niche management', *Technology Analysis and Strategic Management*, **10**, 175–95.
Kemp, R., F. Avelino and N. Bressers (2011), 'Transition management as a model for sustainable mobility', *Transporti Europei*, **47**, 25–46.
Law, J. (2004), 'And if the global were small and noncoherent? Method, complexity, and the baroque', *Environment and Planning D: Society and Space*, **22**, 13–26.
Lawhon, M. and J.T. Murphy (2012), 'Socio-technical regimes and sustainability transitions: insights from political ecology', *Progress in Human Geography*, **36**, 354–78.
Loorbach, D. (2010), 'Transition management for sustainable development: a prescriptive, complexity-based governance framework', *Governance*, **23**, 161–83.

Loorbach, D. and J. Rotmans (2010), 'The practice of transition management: examples and lessons from four distinct cases', *Futures*, **42**, 237–46.

Lyons, G., J. Jain, V. Mitchell and A. May (2012), 'The emergent role of user innovation', in F.W. Geels, R. Kemp, G. Dudley and G. Lyons (eds), *Automobility in Transition? A Socio-technical Analysis of Sustainable Transport*, New York: Routledge, pp. 268–85.

Nelson, R.R. and S.G. Winter (1982). *An Evolutionary Theory of Economic Change*, Cambridge, MA: Belknap Press.

Nill, J. and R. Kemp (2009), 'Evolutionary approaches for sustainable innovation policies: from niche to paradigm?', *Research Policy*, **38**, 668–80.

Nykvist, B. and L. Whitmarsh (2008), 'A multi-level analysis of sustainable mobility transitions: niche development in the UK and Sweden', *Technological Forecasting and Social Change*, **75**, 1373–87.

Raven, R.P.J.M. and F.W. Geels (2010), 'Socio-cognitive evolution in niche development: comparative analysis of biogas development in Denmark and the Netherlands (1973–2004)', *Technovation*, **30**, 87–99.

Reckwitz, A. (2002), 'Toward a theory of social practices: a development in culturalist theorizing', *European Journal of Social Theory*, **5**, 243–63.

Rip, A. and R. Kemp (1998), 'Technological change', in S. Raynor and E.L. Malone (eds), *Human Choice and Climate Change – Resources and Technology*, Columbus, OH: Batelle, pp. 327–99.

Rotmans, J., R. Kemp and M. van Asselt (2001), 'More evolution than revolution: transition management in public policy', *Foresight*, **3**, 15–37.

Schatzki, T.R. (2009), *The Timespace of Human Activity: On Performance, Society, and History as Indeterminate Teleological Events*, Lanham, MD: Lexington Books.

Schwanen, T. (2011), 'Car use and gender: the case of dual-earner families in Utrecht, the Netherlands', in K. Lucas, E. Blumenberg and R. Weinberger (eds), *Auto Motives: Understanding Car Use Behaviours*, Bingley: Emerald, pp. 151–71.

Schwanen, T., D. Banister and J. Anable (2011), 'Scientific research about climate change mitigation in transport: a critical review', *Transportation Research Part A*, **45**, 993–1006.

Schwanen, T., D. Banister and J. Anable (2012), 'Rethinking habits and their role in behaviour change: the case of low-carbon mobility', *Journal of Transport Geography*, **24**, 522–32.

Shove, E. (2012), 'Energy transitions in practice: the case of global indoor climate change', in G. Verbong and D. Loorbach (eds), *Governing the Energy Transition*, London: Routledge, pp. 51–74.

Shove, E. and G. Walker (2007), 'CAUTION! Transitions ahead: politics, practice, and sustainable transition management', *Environment and Planning A*, **39**, 763–70.

Shove, E. and G. Walker (2010), 'Governing transitions in the sustainability of everyday life', *Research Policy*, **39**, 471–6.

Truffer, B. and L. Coenen (2012), 'Environmental innovation and sustainability transitions in regional studies', *Regional Studies*, **46**, 1–21.

Tyfield, D., J. Jin and T. Rooker (2010), *Game-changing China: Lessons from China about Disruptive Low Carbon Innovation*, London: NESTA.

Unruh, G.C. (2000), 'Understanding carbon lock-in', *Energy Policy*, **28**, 817–30.

Urry, J. (2004), 'The system of automobility', *Theory, Culture and Society*, **21**, 25–39.

Urry, J. (2007), *Mobilities*, Cambridge: Polity.
Urry, J. (2011), *Climate Change & Society*, Cambridge: Polity.
Van Bree, B., G.P.J. Verbong and, G.J. Kramer (2010), 'A multi-level perspective on the introduction of hydrogen and battery-electric vehicles', *Technological Forecasting and Social Change*, **77**, 529–40.
Van Driel, H. and J. Schot (2005), 'Radical innovation as a multi-level process: introducing floating grain elevators in the port of Rotterdam', *Technology and Culture*, **46**, 51–76.
Voß, J.-P., A. Smith and J. Grin (2009), 'Designing long-term policy: rethinking transition management', *Policy Sciences*, **42**, 275–302.
Watson, M. (2012), 'How theories of practice can inform transition to a decarbonised transport system', *Journal of Transport Geography*, **24**, 488–96.
Zapata, C. and P. Nieuwenhuis (2010), 'Exploring innovation in the automotive industry: new technologies for cleaner cars', *Journal for Cleaner Production*, **18**, 14–20.

# 15. City transport in a post carbon society

**David Banister**

## INTRODUCTION

The benefits of transport have been central to trade and globalization, and to new forms of networking between people and societies. But all of these activities have led to longer distances and more use of carbon-based fuels (Banister, 2011). While other sectors of the economy have decarbonized, transport has continued to increase its consumption of energy and its carbon dioxide ($CO_2$) emissions. It now accounts for over 25 per cent[1] of global $CO_2$ and this figure is expected to increase to 50 per cent by 2030 (on 2005 levels), as other sectors decarbonize and as transport emissions continue to grow (NEAA, 2009; IEA, 2010).

Global reduction targets have been set, suggesting that a 50 per cent reduction is needed by 2050 (on 1990 levels) in transport to avoid a 2°C increase in global temperatures and sea level rise[2] (ITPS, 2011), and this means that the richer countries should be targeting an 80–90 per cent reduction over this period. Reaching such a reduction target is extremely difficult, but not impossible. At least in cities, there are clear means to create a transport system that uses very little carbon. For longer distance travel the options are far less apparent, and so available oil could be restricted to international aviation and shipping, and the carbon created could be offset through investment in clean energy projects.

Apart from the $CO_2$ question, there are other important reasons as to why change is needed. Transport depends on oil for about 95 per cent of its energy, and since 2008, oil availability has been in decline (Gilbert and Perl, 2010; IEA, 2010). Global demand level for oil is increasing with the growth in motorization in the rapidly expanding economies of the BRIC countries (Brazil, Russia, India and China), and in other countries, but there are also increasing concerns over the high costs of oil (currently about $120 a barrel for Brent Crude), the price volatility and the security of supply. Many countries also subsidize oil to reduce prices on the domestic market, but this distorts the markets and makes alternative energy sources less attractive. Yet the burden on these countries in terms of their levels of debt and the interest payments that need to be made on these imports means that there is less public sector investment available for other needed projects. In 2010 governments spent $250 billion on holding down the prices of petrol and domestic fuel, more than four times the $57 billion they devoted to subsidizing renewable energy. For example, Indonesia's national budget for 2011 sets aside Rp98 200 billion ($11.4 billion) for subsidizing domestic energy, only marginally below last year's Rp99 900 billion. The total is almost four times greater than the Rp26 200 billion that Indonesia will spend on health (Financial Times, 2011). The world needs to break its addiction to oil.

New sources of oil are being discovered in deep waters (for example, off Brazil), in wilderness areas (for example, in the Arctic) and in oil shale and tar sands (for example, in Canada). But in all of these cases the risks and costs are high, and the impacts on the natural environment and the use of energy (and water) for extraction means that the carbon costs are also substantial. Although new supplies of carbon-based energy may help, they only delay the time when society must look beyond the car.

This chapter is really about the future of the car, at least in its present form, and the options for low carbon city transport. At present (2012) there are about 850 million cars and light trucks globally, and this figure is likely to triple by 2050 to 2500 million (OECD and ITF, 2011). Cars and trucks are a global business, involving the motor manufacturers, the oil industry, many suppliers, energy providers, the construction business, support and maintenance providers, and others. This in turn means that they play a substantial role for economies in providing employment, allowing the distribution of goods and services, and in its extension to many parts of everyday life. The transport business is truly global. In addition, there are fragmented patterns of demand, so it is difficult to provide a single alternative, and there is inertia in the system including path dependence and lock-in (Schwanen et al., 2011). It is very difficult to move from one well-established system to another, even if the alternative is well defined and provides a much better way of moving around. To tackle some of these issues, this chapter presents some ideas about the new vehicle system, but suggests that this approach will not tackle the environmental costs of transport and the inequalities created by differential access to transport. It presents a different view of the symbolic icon of the twenty-first century through two different visions of the role of the car in the city, essentially making the case that societies have to move beyond the car, as it is no longer fit for purpose.

## AFTER THE CAR

In their book *After the Car* Dennis and Urry (2009) make a strong case for the car as an iconic emblem of the twenty-first century, and they conclude that in its current form it is outdated. Four points are made, namely, that it is heavy (over 1 tonne), it uses carbon-based fuels (petrol and diesel), it is designed to carry four or more people and it is owned, and they conclude that there should be a new socio-technical system. This post car system would have eight elements, four technological and four socio-organizational. There is clarity on the technological changes needed as these relate to the fuel system (low carbon), new materials (lightweight), smart vehicles (using information and communications technologies (ICT)) and digitization (personal rapid transit (PRT)). On the socio-organizational change, the four elements are less well defined. They include de-privatizing vehicles (new ownership patterns), new transport policies (buses and pricing), new live/work/leisure practices (urban densities) and disruptive innovation (naturally less well specified). In combination, this would provide individualized flexible mobility appropriate for the current century.

It is unclear whether such a change would create a vehicle that effectively counters the high costs associated with the current car, or whether the new system would be low or zero carbon. Six different elements of the current car system are discussed by Dennis and Urry (2009, pp. 36–8) that illustrate its central societal role as a 'way of life and not just as a means of transport from one place to another' (p. 35). They talk about the links between the different business sectors and the iconic names associated with car production over the last 100 years; the car is a major item of household expenditure (after the home) and it has more than just a functional role as it is a sign of adulthood and success, a symbol of status and achievement, as well as portraying notions of freedom and the ideals associated with libertarianism; it supports extensive networks of institutions and organizations, including enforcement, licensing, support, safety, taxation, advertising and marketing agencies, and even urban design and planning; the car culture represents the good life and the development of individualism, exploration and experimentation; it acts as a mobile home, as a detachable part of your property that provides a sanctuary as you pass through strange and dangerous environments in relative comfort and convenience. 'Although automobility is a system of mobility, it necessitates minimal movement once strapped into the driving seat' (Dennis and Urry, 2009, pp. 37–8); finally there are the environmental and resource costs associated with the car, with roads becoming the 'killing fields' of modern society. The question is raised about how such a vehicle has become so dominant and embedded as an essential part of everyday life.

This retrospective view of dominance needs to be reinterpreted in the current societal framework that places a much greater emphasis on sustainability, and the central requirement to reduce levels of resource consumption and emissions. As noted in the introduction, transport has made no contribution to energy reduction and to $CO_2$ emissions reductions, as any efficiency gains have been more than outweighed by growth in demand. The costs to society are substantial, as urban road crashes in the UK cost £12 billion a year (2009 prices), and it affects the young, those living in deprived areas and pedestrians disproportionately, with 62 per cent of UK deaths and serious injuries occurring in urban areas (Cabinet Office, 2009). Air pollutants from transport lead to short- and long-term ill health, and the estimated cost of particulate pollution alone exceeds £9 billion annually (2009 prices) (DEFRA, 2007). Noise causes sleep disturbance and stress, and this may be linked to heart disease (Banister and Woodcock, 2012). Physical inactivity is a risk factor for many diseases, and the estimated costs of sedentary car-based lifestyles and obesity are in excess of £10 billion in England alone (NHS, 2007).

The benefits of car driving are not available to all society, as there are considerable inequalities in access to the car. Almost 30 per cent of the population in the UK will not have access to the car as a driver, due to limitations of age, income, disability or choice, and the costs of learning to drive, insurance and fuel now make it less attractive to young people (Banister, 2009). In addition, the use of the car is heavily skewed towards those in the higher income groups (Table 15.1).

The low-income groups make most use of walk and cycle, and the bus,

*Table 15.1 Average distance travelled by mode and household income: Great Britain, 2010*

| | Lowest income | Second level | Third level | Fourth level | Highest income | All income levels |
|---|---|---|---|---|---|---|
| Walk and cycle | 376 | 325 | 323 | 358 | 378 | 354 |
| Car and van | 4235 | 6240 | 8152 | 9958 | 13322 | 8410 |
| Bus | 795 | 890 | 594 | 446 | 266 | 592 |
| Rail | 451 | 515 | 781 | 992 | 1893 | 926 |
| Other | 295 | 444 | 451 | 588 | 629 | 780 |
| Total | 6152 | 8414 | 10301 | 12342 | 16488 | 10762 |

*Note:* Distances given in km per person per year by mode for the five income quintiles. Other includes motorcycles, private hire buses, taxis, light rail, air and ferries.

*Source:* National Travel Survey (2010), Table NTS0705.

but rail use and car use increases with income. The highest income quintile travels by rail four times further than the lowest quintile, and the ratio for car distance is more than three times. It is the highest income groups that travel much further than the lowest income groups. As with much of transport, carbon-dependent mobility is undertaken mainly by the rich, whether by road, rail or air. About half the UK adult population travels by air in any given year, and this has remained constant (2003–09), and the likelihood of flying is closely related to income (CCC, 2009). Of those with an income level under about £20 000 (57 per cent of the adult population), about 60 per cent do not fly at all, with a further 24 per cent making only one return flight a year. For the richest 22 per cent of the adult population (over £31 000), about 20 per cent do not fly at all, and 45 per cent travel on three or more return flights each year (CCC, 2009). The rich have been the main beneficiaries of the high mobility society that has developed over the last 50 years.

In addition to these environmental costs and inequalities in access is the heavy dependence on the consumption-based lifestyle. Any change towards the post carbon society is reliant on a reassessment of the buy-use-trash culture, and a fundamental change towards one that is based on substantially lower levels of consumption and more on sharing, renting and borrowing. The car itself is a compromise and not good at doing what it ought to do, as it has to fulfil many different functions. Surely, it would be better to have a system that allowed the appropriate vehicle to be used for each different activity, for example, a small vehicle for the school run or shopping, a zero emissions vehicle for city use? This is consistent with Jackson's (2011) case for prosperity without growth, and where intrinsic values are seen to be more important than materialistic values.

There are a substantial amount of resources and public finance embedded within the transport system needed to support the car. This investment has to be paid for, as do the costs imposed on society through the environmental and social costs mentioned above. Alternatives need to be reviewed within a wider framework that considers the carbon costs, the environmental and social factors, the public health implications, as well as the conventional costs and the benefits.

Perhaps the symbolic icon of the twenty-first century is not the car, but the smart phone, as this is what now defines lifestyle through new forms of culture and experimentation, and it allows networking in real and virtual worlds. Its carbon footprint is still substantial,[3] but the smart phone has the functionality to support mobility and to organize the 'best' option for any particular activity in real time and space. This flexibility should be harnessed to reduce and replace the need to travel, rather than to enhance and extend that mobility.

## CHANGING THE DYNAMIC

It is in cities that many positive changes are taking place and the role of the car is being questioned, and in the post carbon society one possible future might be a city with no cars. Many European cities have high levels of walking and cycling with these two modes accounting for about 50 per cent of all trips (Table 15.2). In addition, public transport accounts for a further 20–30 per cent of trips, so the levels of car use are already at low levels, and in some cases are actually reducing (as in Central London).

In developing countries there have been some inspirational examples. In Colombia, Enrique Penalosa has promoted development through the notion of quality of life, interpreted as wellbeing and equality of opportunity. In the capital, Bogota, he has promoted land banking with the public sector owning the land, and controlling its release for different types of activity, and he has raised fuel taxes to pay for investment in pedestrian and cycle networks, and in high quality bus routes (including bus rapid transit). The quality aspect relates to the infrastructure and to the introduction of open spaces, parks and greenways within the city, together with wide pavements (sidewalks). Apart from the physical infrastructure, the key elements are leadership and continuity of actions, and in changing the dynamic within the city so that it is clear as to whom the space belongs. By empowering people and low carbon forms of transport (walk, cycle and

*Table 15.2 Trips by mode in selected European cities*

| City | Foot and cycle (%) | Public transport (%) | Car (%) | Inhabitants |
|---|---|---|---|---|
| Amsterdam (NL) | 47 | 16 | 34 | 718 000 |
| Groningen (NL) | 58 | 6 | 36 | 170 000 |
| Delft (NL) | 49 | 7 | 40 | 93 000 |
| Copenhagen (DK) | 47 | 20 | 33 | 562 000 |
| Arhus (DK) | 32 | 15 | 51 | 280 000 |
| Odense (DK) | 34 | 8 | 57 | 1 983 000 |
| Barcelona (Spain) | 32 | 39 | 29 | 1 643 000 |
| L'Hospitalet (Spain) | 35 | 36 | 28 | 273 000 |
| Mataro (Spain) | 48 | 8 | 43 | 102 000 |
| Vitoria (Spain) | 66 | 16 | 17 | 215 000 |
| Brussels (BE) | 10 | 26 | 54 | 952 000 |
| Gent (BE) | 17 | 17 | 56 | 226 000 |
| Brujas (BE) | 27 | 11 | 53 | 116 000 |

*Source:* ADONIS (1998).

public transport), this has transformed the ownership of the urban realm and people have become more engaged, and they begin to feel proud of their local neighbourhood, rather than being threatened by traffic.

But it is not just retrofitting existing cities with low carbon transport systems that allow high levels of access to all residents and visitors. Masdar provides an example of a demonstration eco-town in a rich Gulf state (Abu Dhabi) with a maxim of 'no carbon no waste'. It is located close to the international airport and covers an area of 6 sq km, and it uses solar power and traditional designs to cool external and internal spaces. It will have a residential population of about 6000 people (growing to 45 000 by 2020), with a further 60 000 commuters travelling by rail from Abu Dhabi. It has been designed by Fosters and Partners,[4] as a traditional walled city with solar panels in the walls, and it is intended to offer a sustainable urban blueprint for the future. There will be no cars in the city, as it is designed at the scale of the pedestrian with shaded paths for walking between buildings. The original intention was to have a PRT system (Podcars) for longer distances within the town that operated automatically between about 100 stops as a driverless system to accommodate an estimated 135 000 trips a day with about 3000 free moving vehicles. This has not yet been implemented, and there is discussion about the feasibility of the Podcars.

## TRANSITION TO A LOW CARBON SOCIETY

There may be two different futures, one based on high technologies that give the same levels of mobility currently provided by the car, but with a much lower carbon profile (Pathway A in Chapter 13). The other is to think about the future of the city in terms of what its functions are and how these might change in the future (the desirable city), and then to think through how low carbon transport might help in achieving that vision (possibly Pathway C in Chapter 13). A third intermediate pathway might follow the compact city model where distances are reduced and there is a greater emphasis on local production and consumption (Pathway B in Chapter 13; OECD, 2012).

Any transition is not easy, particularly when transport and the car are involved. In terms of the former perspective, Dennis and Urry (2009, pp. 156–7) give a graphic description of a post car system, based on the high technology transport future:

> there would be multiple, dense forms of movement of small, ultra-light, smart, probably battery or hydrogen based, deprivatised 'vehicles'. Flexibilised travelling would involve accessing such small, light mobile pods as and when

> required. Electronic regulators embedded in lampposts and in vehicles would regulate access, organise price and control vehicle speed. Some such vehicles would be driverless. The movement of vehicles would be electronically and physically integrated with other forms of mobility. . . There would be a mixed flow of those slow moving micro-cars, as well as bikes, hybrid vehicles, pedestrians and mass transport. These are integrated into networks of physical and virtual access. There would be electronic coordination between motorised and non-motorised transport and between those 'on the move' in many different ways. Smart 'cards' would control access and pay for peoples' uses of the various forms of mobility. And software systems will 'intelligently' work out the best means of doing tasks, meeting up or getting to some place or event.

The transport modes would essentially be publicly owned, there would be integration between the different options, and at the neighbourhood levels proximity through planning for closeness would help reduce travel distances. Pricing and carbon allowances are also a key element in constraining physical mobility, and it is intended that many activities could be replaced by virtual mobility.

The alternative would be to integrate transport as part of the low carbon city that cuts across all sectors (industry, residential, commercial and transport) to ensure efficient use of all forms of energy. Apart from the underlying rationales for the city, in terms of its functions, competitiveness and attractiveness, views on its structure, including the location and density of development, the mixing of different land uses, the amount of open and green space and the design of neighbourhoods – all these factors would constitute essential elements of that debate, with accessibility and proximity being the major transport dimensions. The importance of good quality public transport is central to the concept of the low carbon city, together with easy interchange from one form of shared transport to another, with local trips being made by walk and cycle. The intention would be to ensure closeness to friends, work and other needed services and facilities, so that the need to travel is reduced and the distances to be travelled can be undertaken by walk, cycle and public transport. There would be no need for the 'new' car-based mobility and quality of life would be enhanced.

In the Mercer (2011) quality of life survey 221 cities were compared on 39 criteria, including safety, education, hygiene, health care, culture, environment, recreation, political-economic stability and public transport. The top ten cities in the list are dominated by European cities (Table 15.3), but it is also interesting to note that the proportion of trips by walk, cycle and public transport is also very high, suggesting perhaps that quality of life may be correlated with lower levels of car use in the city.

The choice is clear, whether to replace the car with a new improved version or to take the opportunity to switch to an alternative vision of

*Table 15.3 Mercer 2011 quality of life and trips by walk, cycle and public transport (%)*

| Mercer ranking 2011 | | Walk | Cycle | Public transport | Total |
|---|---|---|---|---|---|
| 1. Vienna | 2010 | 28 | 5 | 36 | 69 |
| 2. Zurich | 2001 | 8 | 5 | 63 | 74 |
| 3. Auckland | 2008 | 13 | 2 | 8 | 23 |
| 4. Munich | 2008 | 28 | 14 | 21 | 63 |
| 5. Dusseldorf | 2004 | 11 | 5 | 31 | 47 |
| 5. Vancouver | 2007 | 17 | 3 | 17 | 37 |
| | | 27 | 3 | 30 | 60 |
| 7. Frankfurt | 2004 | 11 | 7 | 39 | 57 |
| 8. Geneva | 2010 | 49 | | 17 | 66 |
| 9. Bern | 2001 | 11 | 11 | 54 | 76 |
| 9. Copenhagen | 2004 | 6 | 36 | 29 | 71 |

*Note:* The dates refer to the transport data in the table. Vancouver has two figures, one for the city and the other (higher) for the downtown, and Geneva has a combined figure for walk and cycle. Note that Dusseldorf and Vancouver are both fifth equal and Bern and Copenhagen are both ninth equal.

*Source:* http://www.mercer.com/press-releases/quality-of-living-report-2011 (accessed 25 May 2012) and various datasets.

the future city. The car regime is very resilient, and there is considerable investment and commitment to it, as its appeal is based on an individualistic approach to the provision of transport that is privately owned and not shared, with independence, private space, identity and speed being the principal attractions. The car is also embedded in a very successful understanding of social success, with issues relating to sustainability, climate change and oil being reduced to the status of minor concerns. The danger of following the technological pathway is that the carbon savings will not be achieved. The alternative view is that the shared modes of transport (including the bicycle) can offer the same levels of mobility, flexibility, comfort and security in the low carbon city, where speeds are lower and there is more opportunity to appreciate the quality of the city.

Underlying the argument here is the proposition that technological solutions to transport problems accentuate the privileged position that car drivers have maintained over other travellers. This dominance has increased over time, and it is the non-car drivers that have been disadvantaged through poorer quality alternatives, higher levels of pollution, lower levels of safety and a generally poorer quality of environment. There seems to be an assumption embedded in the technological future that more environmentally benign forms of energy will replace oil. The economic, environmental and social arguments would all seem to point towards replacing the car, not with a new improved car, but with a combination of transport that relates more closely to the quality of life in the city.

## CONCLUSIONS

In discussing post carbon mobility, it is not just a question of what can be done, but how such a transition can take place. Giddens (2009) has stated that fear is not the best way to induce low carbon futures, and that there needs to be a positive alternative and new path dependencies. Arthur (2009) suggests that innovation takes 30–40 years and depends on new combinations of elements working together synchronously to create new habits and imaginations. These two views do not suggest that either of the pathways above would be easy, but that there may be some commonalities between them, even if one can be seen as a high technology future (attractive) and the other a low technology future (unattractive).

Part of the solution must lie in a better understanding of the importance of time and how people want to use time in travel and other activities. The concerns over speed and travel time savings need to be related to reliable and reasonable travel times (Banister, 2011). The quality of life in cities is also central, as signified by the designation of clear zones and open spaces for people to meet and spend time together, and as high levels of local pollutants have a direct effect on health. The promotion of active transport also promotes a healthy lifestyle. The reallocation of space in cities to cyclists and walkers with their own separate green networks of routes would create the most suitable conditions for change. By altering the speed and scale of activity, many spaces would become places that people would want to spend time in and money on.

At a more fundamental level, the solution is moving beyond the concern about what can be done with the physical environment to actually improving the conditions for low carbon movement. It is engaging with citizens and talking about a flexible, comfortable, secure personal mobility that is safe and allows travel at a reasonable speed. This is the bicycle in all its manifestations, and it is available to almost all people. The task is to raise the profile of the humble bicycle in terms of the aesthetic and emotional commitment to it, and to think about the means to make it fashionable, faddish and the 'must have' icon. The more fun and cool bike is the obvious solution to the low carbon city transport mode of today and tomorrow.

# NOTES

1. Transport is the second leading source of $CO_2$ emissions, contributing about 7.5 gigatonnes to the 30.6 gigatonnes (Gt) emitted in total in 2010.
2. $CO_2$ concentrations are currently about 385 parts per million by volume (ppmv) and with other greenhouse gases being included, this value increases to 430 ppmv $CO_2e$ (2009). A level of 450 ppmv $CO_2e$ means that there is a 26–78 per cent chance of exceeding 2 C. A level of 550 ppmv $CO_2e$ means that there is a 63–99 per cent chance of exceeding 2°C. If global concentrations of greenhouse gases stabilize at 560 ppmv $CO_2e$, it is 'likely' that global mean surface temperatures will rise by 3 C above pre-industrial levels (Meinhausen, 2006).
3. The Berkeley study (2011) by Ma and Raghavan has estimated the energy used and the embodied energy in computers and phones. This calculation included 750 million desktops and laptops; 1 billion smartphones; 100 million servers; 1 million routers and router-like devices; 100 million local area network (LAN) devices; 5 million cell towers; 75 million telecom switches; 1.5 billion km of fibre optic cabling; and 3.5 billion km of copper cabling for global telecommunications. The estimates on embodied energy of devices are based on previous studies. The team concluded that the internet consumed between 170 and 307 GW, which is equal to between 1.1 per cent and 1.9 per cent of the 16 TW used by humans worldwide. Embodied power was responsible for about 53 per cent of that total.
4. http://www.worldarchitecturenews.com/index.php?fuseaction=wanappln.projectview&upload_id=19047 (accessed 25 April 2012).

# REFERENCES


ADONIS (1998), *Analysis and Development of New Insight into Substitution of Short Car Trips by Cycling and Walking*, Summary report from the FP4 project, available at http://www.vejdirektoratet.dk/dokument.asp?page=document&objno=7134 (accessed 25 May 2012).

Arthur, B. (2009), *The Nature of Technology: What it is and How it Evolves*, New York: Free Press.

Banister, D. (2009), 'Rural transport', in R. Kitchin and N. Thrift (eds), *International Encyclopedia of Human Geography*, Oxford: Elsevier, vol. 11, pp. 460–4.

Banister, D. (2011), 'The trilogy of distance, speed and time', *Journal of Transport Geography*, **19**(4), 950–9.

Banister, D. and J. Woodcock (2012), 'Sustainable urban transport: integrating environmental, health and equity objectives', in M. Barrera and J. Heymann (eds), *Ensuring a Sustainable Future: Making Progress on Environment and Equity*, Oxford: Oxford University Press.

Cabinet Office (2009), *The Future of Urban Transport*, London: The Cabinet Office Strategy Unit and the Department for Transport, p. 27.

CCC (Climate Change Committee) (2009), *Meeting the UK Aviation Target: Options for Reducing Emissions to 2050*, London: Climate Change Committee, December.

DEFRA (Department for the Environment, Food and Rural Affairs) (2007), *Air Quality Strategy*, Cmnd 7169, London: The Stationery Office.

Dennis, K. and J. Urry (2009), *After the Car*, Cambridge: Polity Press.

*Financial Times* (2011), 12 May, available at http://www.ft.com/cms/s/0/b0f56472-7cba-11e0-994d-00144feabdc0.html#ixzz1PRpVKij8 (accessed 25 May 2012).
Giddens, A. (2009), *The Politics of Climate Change*, Cambridge: Polity Press.
Gilbert, R. and A. Perl (2010), *The Transformation of Travel and Freight Movement in a Post-Peak-Oil World*, Vancouver: New Society Publishers.
IEA (International Energy Agency) (2010), 'Transport, energy and $CO_2$: moving towards sustainability', Working Paper, IEA, Paris.
ITPS (Institute for Transport Policy Research) (2011), *Final Report – Transport Systems in a Low Carbon Society*, Final report to the Nippon Foundation, Tokyo, February, available at http://www.jterc.or.jp/english/itps/index.html.
Jackson, T. (2011), *Prosperity Without Growth: Economics for a Finite Planet*, London: Earthscan.
Ma, J. and B. Raghavan (2011), 'The energy and the emergy of the internet', Paper published by ICSI and the University of California at Berkeley, available at http://www.cs.berkeley.edu/~jtma/papers/emergy-hotnets2011.pdf (accessed 25 May 2012).
Meinhausen, M. (2006), 'What does a 2C target mean for greenhouse gas concentrations? A brief analysis based on multi-gas emission pathways and several climate sensitivity uncertainty estimates', in H.J. Schellnhuber, W. Cramer, N. Nakicenovic, T. Wigley and G. Gohe (eds), *Avoiding Dangerous Climate Change*, Cambridge: Cambridge University Press, pp. 253–79.
Mercer (2011), *Worldwide Quality of Living Survey*, 29 November, available at http://www.mercer.com/qualityofliving (accessed 25 September 2012).
National Travel Survey (2010), *Travel by Car Availability, Income, Ethnic Group, Household Size and NS-SEC*, London: Department for Transport.
NEAA (Netherlands Environmental Assessment Agency) (2009), *Getting into the Right Lane*, Report with the Stockholm Resilience Centre and Stockholm University, Bilthoven, October.
NHS (National Health Service) (2007), Health Survey for England 2007. Healthly Lifestyles: Knowledge, Attitudes and Behaviour, London: NHS.
OECD (2012), *Compact City Policies: A Comparative Assessment*, Paris: Organisation for Economic Co-operation and Development, June.
OECD and ITF (2011), *Transport Outlook: Meeting the Needs of 9 Billion People*, Paris: Organisation for Economic Co-operation and Development and International Transport Forum, May.
Schwanen, T., D. Banister and J. Anable (2011), 'Scientific research about climate change mitigation in transport: a critical review', *Transportation Research A*, **45**, 993–1006.

# 16. Thinking change and changing thinking

**David Banister, Moshe Givoni, James Macmillen and Tim Schwanen**

## THE NEED FOR CHANGE IN TRANSPORT THINKING

The need, urgency and benefits of moving to low carbon mobility are clear, and amount to more than the need to reduce carbon emissions, or other environmental harms from transport. This has been highlighted throughout the book. The prospects of anthropogenic climate change and the actions required in this respect only reinforce the need for change in the mobility and transport system.[1] This need has been recognized and for some time, but progress towards low carbon mobility has so far been too slow, in places non-existent. It is commonly stated that carbon reductions in transport are difficult and expensive, suggesting other sectors should curtail emissions. This is not acceptable, especially if reductions elsewhere are offset by increased mobility. So far, transport has taken a low profile when it comes to making a substantial contribution to energy and carbon reduction targets and while this has allowed mobility to increase it has not necessarily favoured society. Moving to low carbon mobility, it is argued in the book, is not only about reducing the costs associated with mobility but also about increasing the benefits from mobility to society. It likewise argues that it is not the amount of mobility but its quality and purpose that matter.

By providing a comprehensive picture, mainly in Part I, on what is happening and what needs to change in the mobility and transport system, the book has laid the foundations for understanding and triggering change. Part II has explored the trajectories of change to low carbon mobility that might be pursued, what low carbon mobility futures might look like and what can be learned from theories of change. This final chapter draws on these two main elements of the book to outline in principle the most important changes that are needed to support and facilitate a meaningful step forward to low carbon mobility.

Primarily, there is a need for a change in 'thinking' when it comes to transport and this is the core message of the book. As long as this thinking (briefly described below) remains unchanged, more research and knowledge of the system, emerging from various research efforts, will not necessarily contribute to moving to low carbon mobility, and will more likely amount to small changes rather than the transition that is required. Any actions to try and move away from high carbon mobility must, first of all, recognize that the mobility and transport system is a complex one, largely built around the main elements that have been covered in Chapters 2–11 but shaped through the various links and feedbacks between them (Chapter 12). Addressing each of the elements alone, and especially the technology element, will not suffice and can be counterproductive. The starting point to advance to low carbon mobility is recognizing that the current high carbon mobility (Chapter 1) is the result of the current links between the whole set of elements in the mobility and transport system, and not due to one of them. The complex nature of the system, while recognized, cannot be used as an excuse for inaction and for a business as usual approach. The way forward this book has suggested is a systemic change and an immediate action to facilitate such change.

While each of the elements of the current mobility and transport system outlined in Chapters 2–11 has a role to play in moving to low carbon mobility, there is another element that weaves all others together, and this is the set of values and norms society adopts or strives to adopt – the current 'lifestyle', in which increasing movement, and thus also mobility, is a defining characteristic. Lifestyle shapes and is shaped by the mobility and transport system, serves to maintain the current situation and might be the most difficult, and important, to change. While addressing lifestyle is beyond the scope of this book, the very much related and associated rationalities – the worldviews, logics and objectivities – and practices of transport policy and governance are at the heart of the book and these need to change (see also Banister et al., 2011).

## THE PROBLEMATICS OF CURRENT TRANSPORT THINKING

Many aspects of current high carbon mobility are derived from two facets of currant transport 'thinking'. First, the way in which travel time is understood and second, the emphasis on infrastructure provision. These are particularly influential rationalities guiding current transport policy that require change as they are major obstacles to low carbon mobility.

In mainstream transport thinking, and following microeconomic theory, travel time is understood as a negative utility or a 'disutility' (see, for instance, Hess et al., 2005). It is only useful insofar as it allows people or freight to reach locations elsewhere but has no (positive) value in and of itself. Travel time tends to be understood in terms of opportunity cost: each minute could be used more productively for other types of activity. For these reasons, transport is considered a derived demand. This reasoning results in prioritizing travel time savings when considering alternative policy options and a valorization of speed. In transport decision-making it is assumed that all else being equal, individuals, households, entrepreneurs and firms will prefer to spend as little (clock) time on travel as possible and/or choose the fastest alternative available. Over the years, a more sophisticated conceptualization of travel time has emerged that recognizes its heterogeneity: travel time varies according to the transport mode used and its attributes, by travel conditions (crowdedness, cleanliness and so on) and by whether it is spent in a vehicle, waiting or as access or egress travel. To account for all these elements together the notion of 'General Travel Time' is sometimes used in recognition of this heterogeneity. Furthermore, these elements of time are often expressed in monetary terms as 'Values of Time', or alternatively together with estimates of the monetary costs of travel combined into some measure of 'generalized' travel costs that ought to be minimized or at least reduced (see, for instance, Tanner, 1979 for an early reference).

From this it follows that providing faster transport connections is a very effective and even 'natural' way to reduce the cost of travel. To get people out of their cars or off a plane and to reduce emissions, there is a need to provide low carbon alternatives that are at least as fast, but preferably (much) faster, than carbon-intensive transport modes. Otherwise, it is assumed, a mode shift will not take place. Reasoning along these lines is a central part of the belief and motivation for investment in, for instance, high-speed trains and bus rapid transport. These two examples also highlight the other facet of mainstream transport thinking, namely, the 'logic of provision' (Schwanen et al., 2011). It is widely taken for granted that the most effective way of intervening in a mobility and transport system is through the provision of transport infrastructure, and that the key decisions in policy-making concern the questions of what type of infrastructure to provide and where. On this view, the mechanism for triggering a shift towards low carbon mobility is to replace investments in roads by expansion of high-quality 'green' infrastructures for freight trains, public transport, cycling and walking, and to concentrate high-density, mixed use urban development at public transport nodes.

The 'logic of provision' also permeates thinking on the congestion 'problem' and ways to solve it. The idea that congestion arises because demand exceeds supply holds true but the notion that increasing supply (of infrastructure) is the solution does not. Although it is recognized that more supply generates more demand (see below), investments in infrastructure continue to be motivated by the desire and perceived need to 'remove bottlenecks', facilitate traffic flow and save the time 'wasted' in traffic. It is this desire that makes congestion charging so desirable in policy circles, even if its effectiveness in solving, or even reducing, congestion can be questioned, as in the case of London (Givoni, 2012). The idea of charging for congestion or other externalities from transport is valid and important. However, the misconception that travel time is merely a disutility means that the expectations regarding the extent to which charging can bring about the required change are generally set too high.

Both the dominance of the idea that travel time is a disutility and the 'logic of provision' are reflective of the history of transport planning. This field has long been dominated by thinking from civil engineering and neoclassical economics and originally came into being to accommodate the growth of car use. In the years after the Second World War a highly technocratic 'predict-and-provide' approach came into being in the US and Western Europe, according to which 'demands [were] projected, equated with need and met by infrastructure provision at least in as far as the public purse [would] allow' (Owens, 1995, p. 44). While arguably much transport infrastructure remains to be built in the Global South, predict-and-provide has largely been abandoned since the 1980s in the North. But despite its demise, some of predict-and provide's ideas and practices remain 'robustly present' (Schatzki, 2010) in contemporary transport thinking and policy. This is, among other things, because most transport models used by national, regional and local planning authorities are direct derivatives of the modelling systems developed under the aegis of predict-and-provide. Transport academics have developed alternative, activity-based modelling approaches[2] but incumbent models, for which travel time is merely disutility, continue to be favoured in most planning organizations. This is because those incumbent models are deeply embedded in the institutional routines and practices of those organizations. There exists, then, a situation of lock-in with regard to transport modelling in practice instead of a level playing field where alternative modelling approaches can displace incumbents.

The principles of predict-and-provide are also embodied in the tools for cost-benefit analysis (CBA) that are currently used for the appraisal of transport investment. Herein travel time reductions often constitute the largest component – often about 80 per cent – of the benefit side of the equation (Banister, 2011). In this way *ex ante* CBA is biased in favour of investments in transport infrastructures that valorize speed and promise to deliver shorter travel times, even though there is hardly any evidence for time savings in the long run. For a long time this situation has favoured investments in road infrastructure and so enhanced the automobility regime (Chapter 14). Nowadays it makes investments in high-speed train attractive.[3]

The prevailing logic with regard to travel time reductions and speed is really too simple and, while not manifested in actual transport policy, this has long been recognized in academic transport research. First, the idea of travel time budgets has been around since the 1970s (Zahavi, 1974; Hupkes, 1982), which holds that at an aggregate level the average travel time is around 60–70 minutes per person per day, irrespective of time, place and culture. Although the idea of stable travel time budgets is contested among academics for a number of reasons (see, for instance, Mokhtarian and Chen, 2004), it has profound implications for mainstream transport rationalities: it implies that populations tend to use faster transport infrastructures not to travel less but to cover greater distances within a roughly similar time budget.

Second, there is an extensive literature on latent or induced demand for transport (Cervero, 2002). Here the idea is that investment in new infrastructure creates new demand, or allows previously unmet demand to become actualized: upon the creation of new infrastructure travelling is made easier and many people (and firms) take advantage of the arisen opportunity to travel farther. After a while the total demand for transport has increased and, so it is often argued, the level of congestion on a given transport network that the initial investment sought to reduce reappears. The idea of, and empirical support for, latent demand is again contested and academic debate continues (Cervero, 2002; Metz, 2008). Nonetheless, the induced demand phenomenon should at least trigger a critical attitude towards the intuitively agreeable notion that investment in infrastructures for low carbon transport would result in a substitution of and shift away from carbon-intensive transport. It is quite conceivable that the newly created infrastructure creates its own demand that is piled on top of existing, carbon-intensive transport patterns. What happens upon new investments in infrastructure is of course an empirical question and the outcomes may vary across regions and situations. Nevertheless, the end result of investment in green infrastructure may be greater mileage overall instead of reduced carbon consumption due to modal substitution. And if, for instance, the creation of a high-speed train network alongside existing aviation systems would mean that air travel is hardly reduced, the end result may even be more greenhouse gas (GHG) emissions than before (Givoni, 2007). This line of argument also invokes scepticism about current claims by politicians and others in the public debate that, post austerity policies, investments in (transport) infrastructures will kill two birds with one stone and both stimulate the economy and make transport greener.

Third, a range of empirical studies have shown that people do not always minimize travel times. Over a series of publications Patricia Mokhtarian and others have argued and shown that travel also has a positive utility (Mokhtarian and Salomon, 2001; Redmond and Mokhtarian, 2001; Lyons et al., 2007), people use travel time to conduct other activities en route – something that is vastly facilitated by the emergence of mobile information and communications technologies, such as mobile phones and tablet personal computers. Even if travel is still on many journeys a disutility, or something the traveller would like to keep within bounds if not avoid, the possibility to use travel time means it is less of a disutility, less undesirable and this means the benefits of saving time, as measured today, are lower. The same research also found that people travel for the sake of travel. This is because it allows them to experience the landscape, speed, control, independence, time of their own and so on. The implication of all this is that travel time is not empty and valueless: people 'use' and experience it in all kinds of ways. In fact, travel time is more than clock time and cannot be reduced to this metric: travel time is lived, and one and the same stretch of clock time is experienced differently across different travellers and by one and the same person across different situations. As a lived reality travel time expands and contracts, flies and grinds to a halt, depending on the context in which it is experienced (see also Bissell, 2008; Watts, 2008).

The implications of these three strands of literature are that transport policies premised on the assumption that people minimize travel time, or optimize this in light of the benefits to be gained at the destination, are likely to be less effective than hoped for and produce unintended consequences. Likewise provision of more, even if different, transport infrastructure is not likely to be effective for low carbon mobility and will likely produce unintended consequences. A shift in thinking that recognizes the problematics in current thinking is a prerequisite for a substantial change in the transport system. There is an urgent need to reframe the transport policy debate and the assumptions in which it is based.

## REFRAMING THE TRANSPORT DEBATE

There is a need to change the logic guiding transport policy and investments in transport, the transport debate must move away from focusing on speed, congestion and infrastructure and be much broader, recognizing the different elements of the transport and mobility system. The argument made is not that travel time reductions and the provision of low carbon infrastructure are not important in moving to low carbon mobility, but rather that they are only part of the story. Access and accessibility need to be understood not only in terms of proximity in physical space, availability in time or physical design barriers but in broader terms (Chapter 3). Access to infrastructure is a necessary but far from sufficient condition for change. As earlier chapters have made clear, carbon-intensive transport is not simply the result of the appropriate infrastructures being in place: there are cultures of automobility and aeromobility in place, as well as vested interests in the realms of the economy and politics (Chapter 14). As Schwanen et al. (2012) argue, changes in people's carbon-intensive travel habits demand much more than changes in transport infrastructure and land use configurations or changes in the monetary price of car transport, although these have a role to play.

The shift in thinking needed is from a preoccupation with the quantity of travel time towards a focus on its quality. This claim does not imply that the amount of travel time is irrelevant but that the idea of 'reasonable travel times' should prevail over their minimization. Reasonable travel time means that travel times should be within bounds – they should be acceptable and not excessive given the context of which they are part. The mentioning of context-specificity is important here, for there are no universal standards or metrics according to which travel times can be evaluated as acceptable or excessive. It is very much a question of what the norms and expectations of different stakeholders are in particular locales and, most important, what passengers are used to in specific places and circumstances. Travel behaviour research shows that people adapt, often quickly, to new travel circumstances and reorganize their life accordingly, when moving on the transport network becomes faster (a new link opened) and similarly when it becomes slower or not possible (when links are closed to maintenance, bridges collapse or a strike stops public transport).

In this respect an element that is often not given enough attention is frequency of travel, as slower journeys might mean the same activities will take place but maybe less often, with some substitute activities taking place instead. Much of the increase in demand for air travel, for example, is the result of people flying more frequently rather than more people flying, which is largely the effect of induced demand as the cost of travel (including time-cost) has been reduced. Not only should travel time be reasonable, there is a need to treat the various elements of the journey differently. One could, for instance, speed up access journeys (to airports, rail stations) but not the flight and train journeys themselves (see

Brons et al., 2009; Givoni and Banister, 2012). Also reasonable should be the frequency of service and this aspect should be considered in tandem with and alongside travel time. While it might be acceptable to provide, and to expect, a bus service to the city centre every five minutes, it is debatable whether more than 50 flights per day between London and Paris or Amsterdam is reasonable, due to both the relative costs and benefits of such a level of service.[4]

With regard to the quality of travel time, it is useful to think in terms of opportunity. Given that one has to travel on a certain mode of transport, what is made possible in that interval? How can the experience of that interval be made as positive as possible? The idea here is that transport policy should make it possible for travellers to use and experience travel time in ways that align with their needs and preferences. Drawing on James Gibson's (1979) notion of affordance[5] is useful in this regard (Watts and Urry, 2008), as it draws attention to what a given environment – a seat in a train carriage, a car on a congested road, a bus stop and so forth – offers to people in terms of opportunities to behave in certain ways or undertake certain practices. By designing the travelling environment in particular ways, it is possible to make time 'fly' – passes quickly and/or in an enjoyable manner – for passengers regardless of the travel time in minutes and hours.

Focusing on affordances means that the spaces travelled in/through should be understood comprehensively and in ways that encompass both the materialities – the 'hardware' of the environment – and the sociabilities – the practices of (other) people, conflicts between such practices, their meanings and associated values and norms. So, the 'walkability' and 'cyclability' of a certain neighbourhood or city are not only about connectivity of street networks, comfort and traffic safety, although these dimensions certainly are very important. Walkability and cyclability concern the total experience and the quality of the encounters with other people, the affective atmospheres, and so on (see also Middleton, 2010). Similarly, train travellers who wish to work on board should be accommodated and equipped not only by the provision of wireless internet connections but also by carriages where the physical design gives them space to work and where they are not disturbed by fellow travellers engaged in other, more noise-producing practices. At the same time, those travelling on a train should not feel themselves overly constrained in their own practices by the needs and preferences of passengers who want to work, sleep and so on. Thinking in terms of quality rather than quantity (of travel time) opens up new ways for developing the unique strengths of low carbon transport modes, which they do not, or hardly, share with their carbon-intensive counterparts. This makes new ways of improving the competitive position of the former possible, and it capitalizes on the idea that the necessity of driving is a key limitation of the car compared to public transport.[6]

As argued in Chapter 9, once transport is expected to play a role in reducing carbon and other emissions, the goal of transport investment is no longer limited to economic growth and increased productivity through facilitation of movement. This, together with the need to treat travel time differently and the need to account for the role of the transport system in issues of equality (Chapters 2, 3 and 13), requires an overhaul in the way transport policies, and investments in infrastructure in particular, are evaluated. As noted, it is likely that until current evaluation practices are changed, the 'wrong' policies will be selected. Yet, changing evaluation practices is not only a matter of changing perceptions and thinking of policy-makers with respect to travel time and the role of transport infrastructure, since quantifying the benefits from investments in quality is often difficult. Making the 'case' for investment in one quality improvement over the other, without quantifying the benefits, is difficult and seen by those needing to justify or defend it as too subjective, thus publicly unacceptable.[7] How to assess policies to improve the quality of the transport system and its services is a real challenge.

## GUIDING PRINCIPLES FOR TRANSPORT POLICY DESIGNED FOR LOW CARBON MOBILITY

The preceding chapters in this book have provided ample suggestions for different policies that can be implemented to advance to low carbon mobility in general and specifically with respect to each of the themes explored. The inventory of measures available for policy-makers is large and diverse enough to promote low carbon mobility. With the same tool box of measures currently available to policy-makers a step towards low carbon mobility can be made. The main issue, or challenge, is which measures to select, in what order and to what extent.

Greater use of pricing, promotion of walking and cycling, integrated transport, integrated land use and transport planning and other policy measures that can contribute to sustainable transport are made regularly, and such measures are increasingly researched, and also tested. There is also a recognition that a host of measures, a policy package,[8] must be implemented in a coordinated fashion for policy interventions to be effective. There is no need to reiterate the importance and role of these in promoting low carbon mobility.

There are, however, policies and measures that do not get enough consideration in the current debate but that might prove particularly important and beneficial, sometimes even at low cost. Suggestions for such policies include the following:[9]

- The cultural meanings and affective atmospheres associated with cars need to be reconfigured, so that popular connotations of cars with freedom, power and control become weakened and the inscription of values and meanings into cars and their use be more realistic and aligned with what the car can actually offer in today's congested traffic.
- More general collective customs bound up with neoliberal capitalism and conducive to car ownership and use, such as the valorization of consumerism and private ownership, need to be challenged. One of the ways in which this can be done is to provide more options for co-ownership and use of cars.
- More restrictions need to be placed on the car industry, so that manufacturers invest more money and energy in non-internal combustion engine-powered cars instead of focusing on incremental changes to existing power-trains.
- Public authorities need to become more consistent in the messages they convey to the general public, businesses and other stakeholders. It is not enough to argue, on the one hand, that low carbon mobility needs to be promoted while, on the other hand, continue to insist on investments in transport infrastructure to promote economic growth.
- From a young age onwards citizens need to be acculturated and trained in the use of low carbon transport modes. They should develop the embodied skills and capacities to use those modes at the time when it is easiest to learn them, so that these competencies can be capitalized on and, if needed, reactivated in subsequent stages of the life-course.
- The governing of the transport system should move away from a focus on single modes of transport and associated networks towards co-modality. Supply of (public) transport services and its governance should be door-to-door not station-to-station or only airport-to-airport. This is already the reality for most freight transport.
- The opportunities afforded by the continuing integration of the physical and the digital realms via social media such as Facebook, Twitter, mobile phone apps and related technologies, as well as the proliferation of digital screens in public and private spaces, need to be harnessed to both reduce mobility and facilitate the use of low carbon modes. This could be done, for example, by allowing full information on the service and in real time, while also allowing the use of travel time in more ways.

- A policy that is hardly ever considered, but is important, is the reduction in the supply of transport infrastructure. In the same way that land was allocated to transport from other purposes it can now be reallocated back, especially in cities and for open, green spaces. When implementing such a policy, a more intelligent and effective use of the remaining capacity needs to be considered. Roads and motorways that are congested tend to be so only in one direction; capacity is often unused on the other side.[10]
- In public transport most efforts (and investments) should be directed at bus transport, both for urban and inter-city travel, and this includes giving buses priority on roads. Efforts must especially focus on addressing buses' main limitation or disadvantage over other modes, their poor image.

## CHANGE BEYOND THE TRANSPORT SYSTEM

While it is evident that major changes in how transport systems are organized and governed are in order, and there is considerable room for action by transport policy-makers, it is also clear that such changes must to a large extent come from 'outside' the transport sector. Transport policy is unable to tackle the challenge of decarbonizing transport on its own and to facilitate the move to low carbon mobility. This is largely because transport policy is derived, first of all, from broader and external factors related to prevailing social values and norms and other policy priorities, like the strive for economic growth (Chapter 13). Furthermore, much decision-making in both the public and the private sectors has transport implications but transport is often not an important consideration, even when the mobility and transport implications are (very) large. Consider, for instance, decisions to build a large hospital, rather than a few smaller ones, and locate this concentrated facility on the outskirts of cities. In short, a systematic change to mobility patterns and demand for transport must be generated from outside the transport sector and transport policy circles.

Transport policy-makers are not 'outside' the mobility and transport system but are an integral part of it and can only act with the current mobility and transport system as a starting point and based on what is currently seen as acceptable, sensible and desirable (Chapter 14). Practices of transport policy-makers must change, but this is unlikely before changes take place outside the mobility and transport system – changes in culture, values and norms leading to a different set of everyday practices. To advance to low carbon mobility there is a need to reassess the buy-use-trash culture and replace it with a culture that is based on substantially lower levels of consumption and more on sharing, renting and borrowing. There is also a need to reassess the spatially and temporally fragmented nature of modern society, which is largely the result of decades of positive feedback loops between private motorized transport ownership and use, urban sprawl, the techno-economic complex centred on the car industry and supportive political and financial institutional arrangement (Chapter 12). New principles such as dematerialization, design for resource sustainability and reusability at any scale are currently being adopted and tried in other policy domains and transport could be integrated into these experimental processes.

It is critically important to move away from the notion that growth can be reduced to increases in gross domestic product (GDP). Consideration ought to be given to factors other than income to define 'improvement' in life, including equality, diversity and the ability of citizens to meet basic and other needs. Wellbeing is often used as shorthand for these broader ways of understanding and appreciating progress (Stiglitz et al., 2009), and the 'capability approach' developed by Amartya Sen, Martha Nussbaum and others offers a particularly useful way of thinking about these issues. The use of the capability approach in relation to mobility and social development – understood broadly across sectors of the economy and domains of society – is still in its infancy (Kronlid, 2008; Beyazit, 2011; Nordbakke and Schwanen, forthcoming). Yet, thinking in the spirit of principles harnessed by this approach is a prerequisite for the emergence of a different approach to governing transport and transport policy.

Even if changes must to a large extent come from 'outside' the transport sector, actors 'within' transport are not absolved from responsibilities and cannot defer these to others. But it does imply that the transition towards low carbon mobility is not primarily the responsibility of public authorities, the users of transport systems or the private parties willing to invest in transport infrastructures. It is a radically distributed, collective challenge in which the business world, the media, politicians and every stakeholder who reproduces the dominance of carbon-intensive transport has to take his or her responsibility. Transition to low carbon mobility is not possible if only the transport sector is trying to change or act as a catalyst for change.

## REALIZING AN OPPORTUNITY FOR CHANGE

There are many signs that the dynamic is changing inside and outside the mobility and transport system and that a window of opportunity for change (Chapter 14) towards low carbon mobility is opening.

Outside transport the economic landscape is unstable and possibly changing, as most economies in the Global North struggle to recover from recent financial crises and seek a return to, from political and policy perspectives, desired economic growth. While policy efforts to return to growth continue, negative growth remains at least for now a stubborn reality in many economies, contributing to increased and continuing unemployment. This 'opens the door' for consideration of alternative models to the current 'economic growth' model – models that do not see investment in transport infrastructure and the facilitation of physical movement (goods and passengers) as an important means to an end (see, for example, Jackson, 2009; Chapter 13). These alternative models are also motivated by concern for the environment, and especially climate change. The financial crises have also brought to the fore concerns about the high cost of living and the increasing gap between the 'rich' and not 'rich'. Although the momentum of 'Occupy' movements has to some extent dissipated, those concerns appear to put further pressure on changing the economic growth paradigm. And this might in turn set in motion a social change that will have mobility implications that are potentially positive from a low carbon point of view. This will be especially likely when the push for equality gathers real pace, given that the benefits of high carbon mobility are disproportionally reaped by the richer strata of society (see, for instance, Brand and Preston, 2010).

Within the mobility and transport system there is some evidence that the current 'automobility regime' is becoming weaker and is changing. In Western countries evidence for saturation of the demand for travel is beginning to emerge, alongside increasing concerns about peak oil and oil prices and in the European Union an official policy target is to: 'Halve the use of "conventionally-fuelled" cars in urban transport by 2030; phase them out in cities by 2050; achieve essentially $CO_2$-free city logistics in major urban centres by 2030' (European Commission, 2011, p. 9). It seems that after the full benefits of the car have been recognized and indeed been capitalized upon, its limitations are now also recognized, especially when oil becomes more expensive and is likely to become ever more expensive in the future. The current forces on and actors within the mobility and transport system show signs of change, which are perhaps best illustrated in the increasing interest in finding alternative fuels and the greater push than before to develop new engines for cars.

Although the signs of change are largely limited to the mature automobility regimes of the Global North, in many respects our mobility system is at, or approaching fast, a crossroad. Insofar as change can be influenced, now is a golden opportunity when it comes to promoting low carbon mobility. As noted in Chapter 13, there are indications that the change that eventually will take place will be predominantly technological with much of the current system and especially the prevailing 'thinking' discussed above remaining largely unchanged. This is, however, unlikely to be sufficient to move to low carbon mobility and would further lock-in the system and makes change in the future even harder to achieve. There is a need to give serious and thorough consideration to pathways to low carbon mobility that go beyond the 'Technological Fix'. Alternative pathways, and especially those where the private car (electric or otherwise) play a relatively limited role in mobility (Chapter 15) and might be harder to enact but offer potentially a much more diverse set of benefits to society over and beyond less carbon consumption must be promoted. Vibrant cities, better public health, less noise pollution, more biodiversity, more social equality – all these goals would then become easier to achieve.

The way forward to low carbon mobility, to sum up, is a full and detailed consideration of all alternative pathways to low carbon mobility and a lot of rethinking when it comes to the transport system and its planning. As suggested by Giddens (2009) with respect to tackling climate change, the low carbon mobility debate needs to focus and emphasize the opportunities and benefits of alternative pathways to low carbon mobility, and less on the risks and disadvantages of the current high carbon mobility.

While the debate is taking place and new approaches to transport planning and governance emerge, develop and disappear, it is vitally important to avoid further lock-in into the current mobility system, especially when it comes to investment in 'heavy' transport infrastructure. Thus, efforts to speed up planning processes and to provide fast-track planning for mega-projects are likely to be counterproductive and wasteful if the 'wrong' infrastructure is put in place. There is a lot of room for actions in transport policy and transport governance as highlighted above, if a change in 'thinking' takes place and investments in new infrastructure and new technologies are relatively low profile in scope and scale.

## NOTES

1. In Chapter 1 mobility is defined as the total amount of travel that is undertaken on all forms of transport and transport is defined as broader than mobility and includes the modes of transport as well as the supply of transport through various institutions and organizations. Transport and mobility are thus closely related and linked, and it is for this reason that the book discusses the 'transport and mobility' system in a holistic way.
2. Some of these have been used in a planning context – Albatross (Arentze and Timmermans, 2004) and FAMOS (Pendyala et al., 2004) are two examples. Many of the activity-based modelling approaches and tools still do not challenge the idea of travel time minimization and incorporate this logic in some way or another.
3. Interestingly, and despite the centrality of speed to aviation, decisions about invest-

ments in air transport infrastructure are not based on travel time savings. Such investments are mostly done on a purely commercial basis and not financed via public resources.

4. On the cost side additional fuel, aircraft and runway capacity can be counted (and the resultant emissions) while on the benefit side mainly the convenience of having a flight at every hour of the day (in economic terms, the reduction in schedule delay).
5. An affordance is a relational and context-specific quality that emerges from the interactions between the capacities of an organism and of an object/setting that allows a certain action to be performed or activity to be undertaken. Consider as an example the interaction between a more-or-less skilled pair of hands and a keyboard that makes typing possible.
6. Although possible to a degree and practised at a considerable scale, many countries ban the use of mobile phones while driving for safety reasons.
7. The perception is that current evaluation practices allow policy-makers to reach, and present to the public, an objective decision.
8. A 'policy package' can be defined as 'a combination of individual policy measures, aimed at addressing one or more policy goals. The package is created in order to improve the impacts of the individual policy measures, minimise possible negative side effects, and/or facilitate interventions' implementation and acceptability' (Givoni et al., 2013).
9. This list is partly based on Schwanen et al. (2012) and Givoni and Banister (2010).
10. 'Intelligent' use of capacity alone will in practice equate to investments in new infrastructure and will likely result in more traffic.

# REFERENCES

Arentze, T.A. and H.J.P. Timmermans (2004), 'A learning-based transportation oriented simulation system', *Transportation Research Part B: Methodological*, **38**(7), 613–33.

Banister, D. (2011), 'The trilogy of distance, speed and time', *Journal of Transport Geography*, **19**(5), 950–9.

Banister, D., K. Anderton, D. Bonilla, M. Givoni and T. Schwanen (2011), 'Transportation and the environment', *Annual Review of Environment and Resources*, **36**, 247–70.

Beyazit, E. (2011), 'Evaluating social justice in transport: lessons to be learned from the capability approach', *Transport Review*, **31**(1), 117–34.

Bissell, D. (2008), 'Visualising everyday geographies: practices of vision through travel-time', *Transactions of the Institute of British Geographers*, **34**(1), 42–60.

Brand, C. and J.M. Preston (2010), '"60-20 emission" – the unequal distribution of greenhouse gas emissions from personal, non-business travel in the UK', *Transport Policy*, **17**(1), 9–19.

Brons, M., M. Givoni and P. Rietveld (2009), 'Access to railway stations and its potential in increasing rail use', *Transportation Research Part A*, **43**, 136–49.

Cervero, R. (2002), 'Induced travel demand: research design, empirical evidence, and normative policies', *Journal of Planning Literature*, **17**(1), 3–20.

European Commission (2011), *Roadmap to a Single European Transport Area – Towards a Competitive and Resource Efficient Transport System*, COM(2011) 144 final, Brussels: European Commission.

Gibson, J.J. (1979), *The Ecological Approach to Visual Perception*, Boston, MA: Houghton Mifflin.

Giddens, A. (2009), *The Politics of Climate Change*, Cambridge: Polity Press.
Givoni, M. (2007), 'Environmental benefits from mode substitution – comparison of the environmental impact from aircraft and high-speed train operation', *International Journal of Sustainable Transport*, **1**(4), 209–30.
Givoni, M. (2012) 'Re-assessing the results of the London congestion charging scheme', *Urban Studies*, **49**(5), 1089–105.
Givoni, M. and D. Banister (eds) (2010), *Integrated Transport: From Policy to Practice*, Abingdon: Routledge.
Givoni, M. and D. Banister (2012), 'Speed: the less important element of the high speed train', *Journal of Transport Geography*, **22**, 306–7.
Givoni, M., J. Macmillen, D. Banister and E. Feitelson (2013), 'From policy measures to policy packages', *Transport Reviews*, **33**(1), 1–20.
Hess, S., M. Bierlaire and J. Polak (2005), 'Estimation of value of travel-time savings using mixed logit models', *Transportation Research Part A: Policy and Practice*, **39**(2–3), 221–36.
Hupkes, G. (1982), 'The law of constant travel time and trip-rates', *Futures*, **14**(1), 38–46.
Jackson, T. (2009), *Prosperity Without Growth: Economics for a Finite Planet*, London: Earthscan.
Kronlid, D. (2008), 'Mobility as capability', in T. Priya Uteng and T. Cresswell (eds), *Gendered Mobility*, Aldershot: Ashgate, pp. 15–34.
Lyons, G., J. Jain and D. Holley (2007), 'The use of travel time by rail passengers in Great Britain', *Transportation Research Part A: Policy and Practice*, **41**(1), 107–20.
Metz, D. (2008), 'The myth of travel time saving', *Transport Reviews*, **28**(3), 321–36.
Middleton, J. (2010), 'Sense and the city: exploring the embodied geographies of urban walking', *Social and Cultural Geography*, **11**(6), 575–96.
Mokhtarian, P.L. and C. Chen (2004), 'TTB or not TTB, that is the question: a review and analysis of the empirical literature on travel time (and money) budgets', *Transportation Research Part A: Policy and Practice*, **38**(9–10), 643–75.
Mokhtarian, P.L. and I. Salomon (2001), 'How derived is the demand for travel? Some conceptual and measurement considerations', *Transportation Research Part A: Policy and Practice*, **35**(8), 695–719.
Nordbakke, S. and T. Schwanen (forthcoming), 'Wellbeing and mobility: a theoretical framework and literature review focusing on older people', *Mobilities*.
Owens, S. (1995), 'From "predict and provide" to "predict and prevent"?: Pricing and planning in transport policy', *Transport Policy*, **2**(1), 43–9.
Pendyala, R.M., R. Kitamura, A. Kikuchi, T. Yamamoto and S. Fujii (2004), 'Florida activity mobility simulator: overview and preliminary validation results', *Transportation Research Record*, **1921**, 123–30.
Redmond, L.S. and P.L. Mokhtarian (2001), 'The positive utility of the commute: modeling ideal commute time and relative desired commute amount', *Transportation*, **28**(2), 179–205.
Schatzki, T.R. (2010), *The Timespace of Human Activity: On Performance, Society, and History as Indeterminate Teleological Events*, Lanham, MD: Lexington Books.
Schwanen, T., D. Banister and J. Anable (2011), 'Scientific research about climate change mitigation in transport: a critical review', *Transportation Research Part A: Policy and Planning*, **45**(10), 993–1006.

Schwanen, T., D. Banister and J. Anable (2012), 'Rethinking habits and their role in behaviour change: the case of low-carbon mobility', *Journal of Transport Geography*, **24**, 522–32.
Stiglitz, J.E., A. Sen and J.-P. Fitoussi (2009), *Report by the Commission on the Measurement of Economic Performance and Social Progress*, available at http://www.stiglitz-sen-fitoussi.fr (accessed 20 June 2012).
Tanner, J.C. (1979), *Expenditure of Time and Money on Travel*, Crowthorne, Berkshire: Transport and Road Reseach Laboratory.
Watts, L. (2008), 'The art and craft of train travel', *Social and Cultural Geography*, **9**(6), 711–26.
Watts, L. and J. Urry (2008), 'Moving methods, travelling times', *Environment and Planning D: Society and Space*, **26**(5), 860–74.
Zahavi, Y. (1974), *Travel Time Budgets and Mobility in Urban Areas*, Washington, DC: Federal Highway Administration.

# Index